Lecture Notes
in Business Information Processing   69

Frank Harmsen   Erik Proper
Frank Schalkwijk   Joseph Barjis
Sietse Overbeek (Eds.)

Practice-Driven Research on Enterprise Transformation

Second Working Conference, PRET 2010
Delft, The Netherlands, November 11, 2010
Proceedings

 Springer

Volume Editors

Frank Harmsen
University of Maastricht
Minderbroedersberg 4-6, 6211 LK Maastricht, The Netherlands
E-mail: f.harmsen@maastrichtuniversity.nl

Erik Proper
Public Research Centre Henri Tudor
29, avenue John F. Kennedy, 1855 Luxembourg-Kirchberg, Luxembourg
E-mail: erik.proper@tudor.lu

Frank Schalkwijk
Atos Origin
Papendorpseweg 93, 3528 BJ Utrecht, The Netherlands
E-mail: frank.schalkwijk@atosorigin.com

Joseph Barjis
Delft University of Technology
Faculty of Technology, Policy and Management
Section of Systems Engineering
P.O. Box 5015, 2600 GA Delft, The Netherlands
E-mail: j.barjis@tudelft.nl

Sietse Overbeek
Delft University of Technology
Faculty of Technology, Policy and Management
Section of Information and Communication Technology
P.O. Box 5015, 2600 GA Delft, The Netherlands
E-mail: s.j.overbeek@tudelft.nl

Library of Congress Control Number: 2010937670

ACM Computing Classification (1998): J.1, H.3.5, H.4.1, D.2, K.6.3

ISSN	1865-1348
ISBN-10	3-642-16769-1 Springer Berlin Heidelberg New York
ISBN-13	978-3-642-16769-0 Springer Berlin Heidelberg New York

springer.com

© Springer-Verlag Berlin Heidelberg 2010
Printed in Germany

Typesetting: Camera-ready by author, data conversion by Scientific Publishing Services, Chennai, India
Printed on acid-free paper   06/3180   5 4 3 2 1 0

Preface

The PRET working conferences are set up as a one-day event in such a way that it attracts an audience from both industry and academia. PRET 2010 was a continuation of the PRET 2009 working conference, which was organized as the industrial track at the 2009 CAiSE conference. The PRET 2010 working conference was organized as part of the enterprise engineering week, and was co-located with PoEM 2010 and TEAR 2010 in Delft.

The statement that modern-day enterprises are in a constant state of flux is in 2010 even more true than it was in 2009. The markets are in a state of confusion and seem to have no direction at all, as they are swinging back and forth depending on often contradictory signals and economic forecasts. As a consequence, enterprises, be they private businesses, government departments or other organizations, are taking their measures. Restructuring, divesting, improving performance and merging are among the usual transformation activities that enterprises conduct to provide answers to the ever-challenging demands that are put on them. In addition to the tricky economic situation, developments like globalization, rapid technological advancement, aging and the changing mindset of customers contribute to a situation in which nothing is certain anymore and in which change is the only constant.

PRET approaches these developments and the impact they have on enterprises from a holistic enterprise engineering perspective. Typical questions that are answered in our working conference are:

- How can information technology support and enable enterprise transformation?
- How can enterprises and their transformation be modeled?
- How are information systems transformation and enterprise transformation related?
- How should a transformation be managed?
- How should a transformation be constructed, given the situation at hand?

In the answers, topics are addressed from the people, the process and the technology perspective, thus creating a balanced mix of these three aspects, which are equally important in enterprise transformation.

An important objective of PRET is the collection of real-life case studies regarding enterprise transformation. This supports our view that the research area of enterprise transformation (or enterprise engineering, for that matter), can only thrive if industry and academia closely cooperate: the latter to develop concepts, paradigms, tools and methods, and the first to validate them in the "real world."

This objective is reflected in the accepted papers of PRET 2010. This year, the Program Committee selected 9 excellent papers from 24 submissions. The authors were asked to combine theory with practice, using real-life case studies and practical experiences. Most authors acted on this request. The one or two papers without real-life experience were admitted because they concerned high-quality research that could benefit from the confrontation with the industry. In that sense, PRET acts as a platform to bridge the gap between theory and practice and to create synergy and cross-fertilization.

Each paper was allocated to one of the three tracks of this working conference:

1. Situational Transformation
2. Portfolio, Program and Project Management
3. Enterprise Architecture to Align Business and IT

The papers are submitted as book chapters, with an average size of about 25 pages. This size, enabling a slightly more in-depth coverage of the research topic, should stimulate meaningful discussion, with the goal of developing the field of enterprise transformation, creating synergy and jointly identifying topics for further research.

August 2010   Frank Harmsen

Organization

Steering Committee

Erik Proper	Public Research Centre Henri Tudor, Luxembourg, and Radboud University Nijmegen, The Netherlands
Frank Harmsen	Ernst & Young IT Advisory and Maastricht University, The Netherlands

Organizing Committee and Program Board

Chair: Frank Harmsen	Ernst & Young IT Advisory and Maastricht University, The Netherlands
Erik Proper	Public Research Centre Henri Tudor, Luxembourg, and Radboud University Nijmegen, The Netherlands
Bas van der Raadt	Ernst & Young IT Advisory, The Netherlands
Frank Schalkwijk	Atos Origin, The Netherlands
Knut Grahlmann	Ernst & Young IT Advisory, The Netherlands
Sietse Overbeek	Delft University of Technology, The Netherlands
Joseph Barjis	Delft University of Technology, The Netherlands

Program Committee

Antonia Albani	Delft University of Technology, The Netherlands
Jan Dietz	Delft University of Technology, The Netherlands
Eric Dubois	Public Research Centre Henri Tudor, Luxembourg
Matthias Ekstedt	Royal Institute of Technology, Sweden
Gregor Engels	Capgemini SDM and University of Paderborn, Germany
Jaap Gordijn	VU University Amsterdam, The Netherlands
Frank Harmsen	Ernst & Young IT Advisory and Maastricht University, The Netherlands
Jan Hoogervorst	Sogeti, The Netherlands
Stijn Hoppenbrouwers	Radboud University Nijmegen, The Netherlands
Marta Indulska	University of Queensland, Australia
Pontus Johnson	Royal Institute of Technology, Sweden
Patricia Lago	VU University Amsterdam, The Netherlands
Marc Lankhorst	Telematica Instituut, The Netherlands

Jan Mendling	Queensland University of Technology, Australia
Hans Mulder	VIA Groep, The Netherlands and
	University of Antwerp, Belgium
Sietse Overbeek	Delft University of Technology, The Netherlands
Anne Persson	University of Skövde, Sweden
Erik Proper	Public Research Centre Henri Tudor,
	Luxembourg, and
	Radboud University Nijmegen, The Netherlands
Hajo Reijers	Eindhoven University of Technology,
	The Netherlands
Pnina Soffer	University of Haifa, Israel
Stefan Strecker	Duisburg-Essen University, Germany
Inge van de Weerd	University of Utrecht, The Netherlands
Bas van der Raadt	Ernst & Young IT Advisory, The Netherlands
Bas van Gils	BiZZdesign, The Netherlands
Johan Versendaal	University of Utrecht, The Netherlands
Robert Winter	University of St. Gallen, Switzerland

Subreviewers

Bram Klievink	Delft University of Technology, The Netherlands
Christian Soltenborn	University of Paderborn, Germany
Hannes Holm	Royal Institute of Technology, Sweden
Markus Buschle	Royal Institute of Technology, Sweden
Ulrik Franke	Royal Institute of Technology, Sweden
Yiwei Gong	Delft University of Technology, The Netherlands

Endorsing Organizations

BeInformed, The Netherlands
Fraunhofer Institute for Applied Information Technology, Germany
Public Research Centre Henri Tudor, Luxembourg
The Netherlands Architecture Forum (NAF), The Netherlands
Dutch Research School for Information and Knowledge System (SIKS),
    The Netherlands
Radboud University Nijmegen, The Netherlands
SeederDeBoer, The Netherlands
Delft University of Technology, The Netherlands

Supporting Organizations

Atos Origin, The Netherlands
Capgemini, The Netherlands
Ernst & Young, The Netherlands

Table of Contents

Situational Transformation

Portfolio, Program and Project Management

Enterprise Architecture to Align Business and IT

Design and Engineering for Situational Transformation

Gerrit Lahrmann[1], Robert Winter[1], and Marco M. Fischer[2]

[1] Institute of Information Management,
University of St. Gallen, Switzerland
{Gerrit.Lahrmann,Robert.Winter}@unisg.ch
[2] SAP Business Transformation Services,
SAP (Switzerland) Inc.
Marco.Fischer@sap.com

Abstract. Enterprise architecture management (EAM) is an instrumental means to increase transparency, consistency, simplicity, flexibility (i.e. the ability to adjust), and/or ultimately agility (i.e. the ability to innovate/transform). With to-be architecture models as well as evolution guidelines being inherently prescriptive, EAM research is predominantly design-oriented. However, the design and engineering of prescriptive artefacts needs to consider the specifics of problem situations. Based on existing classifications for EAM approaches and for transformation project situations, the potential contributions of EAM in transformation projects are analyzed. Based on this overall analysis, a concrete transformation situation is selected and further analyzed. Cluster analysis of survey data is applied to identify impacts and design configurations as an exemplary problem analysis for the specification of EAM-enabled transformation.

Keywords: enterprise architecture, enterprise engineering, organizational design, organizational engineering, transformation.

1 Introduction

Enterprise Architecture Management (EAM) can be a means for a broad range of ends: Depending on an organization's needs, it can be instrumental to increase transparency, consistency, simplicity, flexibility (i.e. the ability to adjust), and/or ultimately agility (i.e. the ability to innovate/transform). With to-be architecture models as well as evolution guidelines being inherently prescriptive, EAM research is predominantly design-oriented.

Since no single solution artefact can be expected to fit all problems of a certain design problem class, the design and engineering of prescriptive artefacts needs however to consider the specifics of the problem at hand. As a compromise between 'one-size-fits-all' solutions whose disadvantage is the missing fit on the one side, and problem specific solutions whose disadvantages are the immense construction effort and their missing generality on the other, usually a handful of problem situations are differentiated which cluster related design problems [1]. A problem situation is specified by a combination of certain contingency factors with certain project goals that imply to differentiate the implied design problems from other, related design problems that have other contingencies or are subject of other goals. Design science research in the

F. Harmsen et al. (Eds.): PRET 2010, LNBIP 69, pp. 1–16, 2010.
© Springer-Verlag Berlin Heidelberg 2010

EAM field needs to build upon a well-understood knowledge base of design/engineering goals and contingencies of 'typical' problem situations. Situation specific solution artefacts are designated as approaches in the following.

For EAM in general, first contributions have been made that identify problem situations and propose situation-specific EAM approaches such as (i) the mature approach, (ii) the rather passive IT-biased approach, and (iii) the initial approach [2]. But this proposal does not relate EAM to transformation explicitly.

On the other hand, there is a growing understanding of transformation situations and respective, situation-specific transformation management approaches [3] such as (i) strategy adaptation, (ii) business networking, (iii) technology enabled growth, and (iv) process redesign – but this classification does not consider EAM. In particular, it is not clear which EAM approach is an appropriate means to achieve the respective, situation-specific transformation goals.

The paper at hand addresses this research gap. Based on existing situational approaches for EAM and for transformation, the potential contributions of EAM in transformation projects are analyzed. Cluster analysis of survey data is applied to identify impacts and design configurations, which are used to further detail representative EAM-enabled transformation situations. The results form the foundation for the construction of situational, and therefore more effective, artefacts like e.g. method fragments, reference models, design principles, and ultimately methods.

The paper is structured as follows. In section 2, we give an overview of our research approach. Related work is discussed in section 3. In section 4, we develop our research model. The research model is evaluated by means of an empirical analysis in section 5. In section 6, we discuss the results of the evaluation. In the concluding section 7, we offer suggestions for future work.

2   Research Overview

In order to analyze the potential contributions of EAM in transformation projects, we aim at the identification of impacts and design configurations of various transformation project drivers. The results form the foundation for the construction of prescriptive artefacts, i.e. constructs, models, methods, or instantiations as "technology-based solutions to important and relevant business problems." [4]

The rigorous construction of useful IS artefacts is typically attributed to the design science research (DSR) paradigm [5]. Design problems in organisations are generically defined as "the differences between a goal state and the current state of a system" [4]. Most authors recommend to start the DSR process with the identification of the important and relevant problem that is going to be addressed, but a concrete methodological support how to identify a design problem, how to show its importance and relevance, and how to understand the design problem sufficiently to support subsequent solution design is missing [1].

Besides of being important and relevant, the design problem – and hence the proposed design solution (= DSR output artefact) – should be sufficiently general [1]. For Hevner et al. [4], generality is one of three quality criteria of an DSR artefact. Baskerville et al. [6] demand a design research artefact to "represent [...] a general solution to a class of problems." Therefore, it can be assumed that DSR results are generic (and not specific) IS artefacts which are useful for solving a class of design problems [1].

The two research goals of generality and utility are conflicting. In their research on reference modelling, Becker et al. [7] designate this trade-off as reference modelling dilemma: "On the one hand, customers will choose a reference model that […] provides the best fit to their individual requirements and therefore implies the least need for changes. On the other hand, a restriction of the generality of the model results in higher turn-over risks because of smaller sales markets" [7]. This dilemma is not only apparent in reference modelling, but also exists for other general solutions to classes of design problems (e.g. methods). As a solution to this dilemma for (reference) models, Becker et al. [7, 8] propose adaptation mechanisms that instantiate a generic reference model according to the specific design problem at hand. Transferred to the general DSR context and referring to all four artefact types identified by March & Smith [5], the extension of generic artefacts by adaptation mechanisms can be designated as situational artefact construction (SAC) [1]. In addition to situational reference modelling [e.g. 7], situational artefact construction has also been investigated for methods (situational method engineering, see e.g. [9]).

As SAC allows the researcher to develop a set of situation-specific artefacts ("the approach") which might be even adaptable to different problems within a design problem class, a crucial decision during the construction phase is to delineate the range of addressed design problems (i.e. to specify the design problem class) and to understand the relevant design situations within this class [1]. Depending on the degree of generality, a design problem class can be partitioned into few, very generic situations or a larger number of (different) situations of lesser generality. Thereby, it also becomes intuitively clear that solution artefacts can be constructed on different levels of generality – the fewer artefacts are to be constructed, the higher their generality has to be [1].

According to Winter [1], a fundamental understanding of the design problem class is needed, which can be broken down into identifying relevant properties ("design factors") of the addressed design problems, specifying the property ranges that imply the addressed design problem class, defining metrics for these properties that represent the similarity/dissimilarity of design problems within that class, calculating the ultrametric distances based on these metrics, and using these ultrametric distances for specifying the desired generality level of the solution artefacts that are to be constructed in subsequent DSR steps. Thereafter, the design problem class has been sufficiently analyzed to allow for the specification of situations and, in a next step, the systematic development of solution artefacts (i.e. a solution approach) for one or more situations.

Concluding, in order to be able to build better artefacts as means to specific problem solutions, our goal within this paper is to better understand typical situations in transformation projects. In order to reach this goal, we proceed as follows. First, we analyze the related work on contingency theory, EAM, and transformation in order to specify situations for EAM-enabled transformation in form of a generic research model. Next, we focus on ERP implementation projects as an exemplary sub-class of EAM-enabled transformation. For that exemplary situation, we identify key drivers of transformation projects on the basis of an empirical study. We conclude by discussing the generalization of our findings to the field of EAM-enabled transformation and by outlining future work.

3 Related Work

Contingency theory, EAM, and transformation are fundamental to understand the research model presented subsequently in section 4. We summarize related work in these areas in the following.

3.1 Contingency Theory

Building on the seminal work of Fiedler [10], who was the first to consider contingency respectively situational factors like the leader's power position in studies on leadership effectiveness, contingency theory's quintessence is that the effect on one variable caused by another depends on a third (or multiple), moderating one(s) [11] (cf. figure 1).

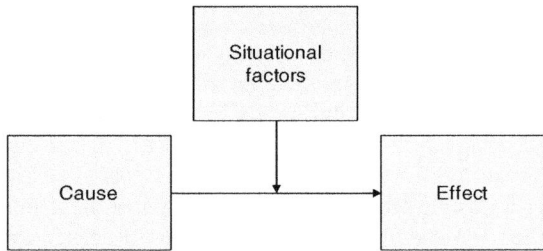

Fig. 1. The gist of contingency theory: situational factors as moderating variables

   Putting this in an organizational context, e.g. transformation projects, often cited situational factors in organisation theory are organizational size, task uncertainty, and task interdependence [12]. These situational factors play an important role, as a higher fit or goodness of alignment of an organizations situation to these factors [13, 14] usually leads to a higher overall success of the organization, whilst a misfit will usually lead to a lowered performance [15].
   Due to the strong dependency of organizational performance on fit, the design and engineering of prescriptive artefacts needs to consider the specifics of problem situations. Nevertheless, it has not been proved empirically that contingency factors from organisation theory are also valid for the design and engineering of prescriptive artefacts – therefore, further research is needed to identify and validate appropriate situational factors [1].

3.2 Enterprise Architecture Management

EAM can be an instrumental means to increase an organizations ability to transform. For EAM, aggregated models need to be created and maintained which cover a broad scope from business artefacts to IT artefacts [16]. Due to the broad nature of the EA field, EA applications and application situations are manifold, still relatively immature in practice, and differ significantly from industry to industry [17]. Although some EA methods take into account the respective application situation [18], it is relatively unclear which situational factors have an effect on their appropriateness. Therefore,

EAM approaches have been presented in the literature, which are derived from observations in practice [2]. The classification is based on a combination of determining factors into statistically relevant clusters. EA methods should consider these 'design factors'. They can be referred to as decisive factors describing the context in which projects transform EA. Therefore, certain typical design factor constellations can be identified which provide insight on how to approach EAM. Aier at al. [2] identify the adoption of advanced architectural design paradigms and modelling capabilities, the deployment and monitoring of EA data and services, and the organizational penetration of EAM as relevant design factors (cf. table 1).

Table 1. EAM design factors [2]

#	Name	Description
1	Adoption of advanced architectural design paradigms and modelling capabilities	This factor describes valuable ways to adopt the concept of EA. On the one hand, it involves well established architecture design paradigms which emphasize the layered structure of EA. On the other hand, this factor makes clear that a further enhancement of EA also depends on the dimension of the EA documentation. To allow for a continuous development, not only loosely coupled artefacts, but also an idea of how to approach a future development stage is decisive. EA then contributes to business/IT alignment by offering simulation capabilities, which presupposes different variants of its to-be structures.
2	Deployment and monitoring of EA data and services	This factor describes the deployment of EA within the organization. It is required to establish a consistent monitoring of EA data and services to further enforce the deployment. This can be assisted by the role of an EA quality manager who is responsible for observing periodic reviews of EA data and EA processes. A high degree of EA deployment puts the organization in the position to reduce its costs for maintenance activities, software and hardware licenses, but also to ensure that similar concerns are treated equally and according to the parameters of the EA roadmap. A high factor value also points to the application of sophisticated EA analysis techniques within the organization.
3	Organizational penetration of EAM	This factor accounts for the penetration of EA in the organization. The overall level of penetration is influenced by the degree EA results and EA documentation are used by a broad range of stakeholders. Therefore, EA is a suitable tool not only to support IT-related work, but also to serve the business units and to provide reliable information to management units. The level of organizational penetration increases with the organization's capability to clearly communicate EA benefits to the potential stakeholders – regardless if they actually operate on EA results or not. Therefore, this factor describes the way EA is perceived and utilized across the organization. A high level of organizational penetration leads to a higher acceptance, and less misinterpretation of EA within the organization, respectively.

Based on these factors, it has been confirmed that there is no overall approach to adapt to EAM in practice, but that three different EAM approaches can be distinguished. Overall, these three EAM approaches represent different interpretations on how to grasp EAM in terms of its determining factors (cf. table 2).

Table 2. EAM approaches [2]

#	Name	Description
1	Mature EAM	In the "Mature EAM" approach, EAM is understood as a valuable instrument to develop and thus transform EA in its holistic understanding. It relies on a progressive perception of EA within the business and management units. Mature EA in its current state in practice has an intermediate maturity regarding the employment and monitoring of EA data and services. Therefore, there still is some development potential in practice and this approach may be interpreted as a not fully developed instantiation of EAM.
2	IT-biased EAM	"IT-biased EAM" is well anchored in the IT domain and has reached an average maturity there. However, this limited architectural understanding is an obstacle in order to really leverage the value of available IT understanding, models, and methods. As regards the design problem, rather advanced architectural design paradigms – e.g. service orientation – are not much developed in this approach because they require a certain amount of organizational penetration.
3	Initial EAM	"Initial EAM" puts emphasis on transparency as the necessary precondition to realize benefits from EA application. Organizations using this approach are in particular interested to implement relevant applications to demonstrate these benefits. This also explains the need for more sophisticated analysis techniques – which organizations using this approach often lack of. This typically is a hint for a tool driven or model driven approach to EA design as opposed to an application driven approach. Such a tool driven approach may be dangerous since it requires significant efforts to survey and model the architectural data without a clear idea of future application situations.

Although the determining factors and approaches provide a basis on which situational EA method constructions should derive their contexts in future, this classification does not consider different EAM goals such as increasing transparency, consistency, simplicity, flexibility and/or agility. In particular, it is not clear which transformation situations call for which EAM approach.

3.3 Transformation Management

According to Baumöl [3], change projects are unique, because they are embedded into unique contexts, i.e. the economic, technological and social environment. As a consequence, the setup and the execution of transformation projects need to refer to this unique context [3]. The basis for this is the construction of adequate interdisciplinary methods which do not only concentrate on one aspect of the transformation process, e.g. the role of IT, but rather include the main levers of organizational change from all

Table 3. Main topics of transformation projects [3]

#	Name	Description
1	Strategy	Strategic agility in the sense of changing the organization was pinpointed to the clarity about today's situation, the existence of a concept how the future organization should look like, and the definition of a transformation process (or strategy process) that focuses on involving the people.
2	Leadership	Establishing role models and with this providing permission for a changed behaviour. Moreover, the assignment of ownership for specific aspects of the required changes. Cultural aspects and their role in change projects form a basis on which change efforts are built, but are no specific drivers or inhibitors in the transformation process.
3	Sustainability	The enduring manifestation of an organization in form of its operational structures (i.e. business processes) and the mid-term success of transformation projects (in addition to the short-term success being on time, in scope, in budget). In comparison to the process architecture, an organization chart is quite volatile. Therefore, business process management is a prime lever for creating a more flexible and nonetheless sustainable organization.
4	Performance measurement	Controlling the change effort and its effects is important for achieving successful organizational change and thus strategic agility. Main factors influencing the performance measurement system are the compatibility of the leading company culture and the metrics used, the underlying attitude towards measuring [19], the transparency of the system and the results for the employees.
5	IT	The role and tasks of IT can be clustered in three main topics, i.e. the support of communication throughout the organization, the creation of strategic agility for the business, and the structuring and support of business processes.

relevant disciplines (e.g. strategy making, organizational design and behaviour, or business process engineering) [3].

In order to enable the construction of such methods, a precondition is to gain an understanding of how to systematically structure and manage transformation projects [3]. Therefore, based on the analysis of 89 interviews, published case studies, and existing methods, Baumöl [3] identifies the leading content-related situational factors of transformation projects and aggregate these factors in form of transformation situations. The major topics to be addressed in strategic transformation projects are summarized in table 3.

The transformation situations which aggregate the aforementioned situational factors of transformation describe recurring frames of the content-related factors with respect to specific situations. Table 4 summarizes the identified transformation situations.

Table 4. Transformation situations [3]

#	Name	Description
1	Strategy adaptation	Change projects having a focus on comprehensive strategy adaptation.
2	Business networking	Change projects having a focus on redesigning the communication and interaction with customers and the business network.
3	Technology enabled growth	Change projects dealing with growth strategies and cultural aspects placed in a technological context.
4	Process redesign	Change projects having a focus on process engineering or process redesign.
5	Agility improvement	Change projects dealing with the improvement of agility of the organization.

4 Research Model

In the following, we present our research model for linking EAM and transformation management in order to specify situations for EAM-enabled transformation.

In section 3.1, we briefly outlined the gist of contingency theory that situational factors act as moderating variables in means-end-chains and that organizations, in order to achieve high organizational performance, need to address these factors by ensuring appropriate fit. An implication of this is that the design and engineering of prescriptive artefacts needs to consider the specifics of problem situations. In section 3.2, we sketched EAM as an instrumental means to increase an organizations ability to transform. Based on Aier at al. [2], we elaborated that the three EAM approaches (and respective problem situations) "Mature EAM", "IT-biased EAM", and "Initial EAM" can be differentiated. On the basis of Baumöl [3], we delineated five different transformation situations in section 3.3. The five transformation situations (and respective management approaches) "Strategy adaptation", "Business networking", "Technology enabled growth", "Process redesign", and "Agility improvement" represent archetypical ends of transformation projects.

Putting all this together, we propose that different transformation situations require different EAM approaches in order to achieve the best possible performance as regards organizational transformation. The selection of the appropriate EAM approach depends on the contingent factors of the transformation situation. Therefore, we want to identify the relations between EAM approaches (means) and transformation situations (end). Figure 2 summarizes this in form of a conceptual research model.

According to Moreton, IT can be an important enabler and integrator in transformation projects by combining the "design, development and exploitation of systems and their organizational context" [20]. Baumöl (cf. section 3) identified this situation as one of five typical transformation situations and termed this transformation situation "Technology enabled growth". A good example of transformation projects of the "Technology enabled growth" situation are highly integrative ERP implementation projects, as they could potentially "change the infrastructure and operating practices

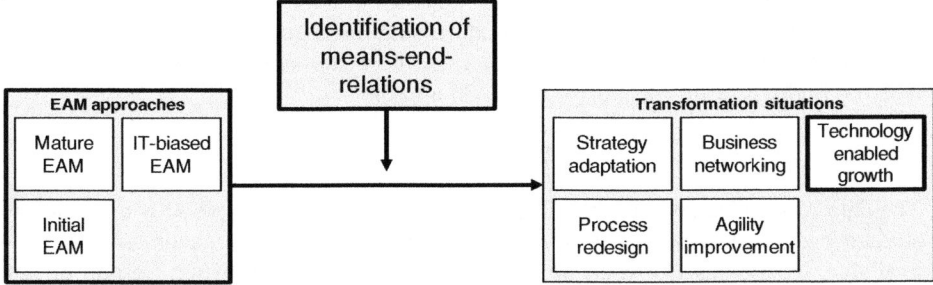

Fig. 2. Research model (focus of the empirical evaluation highlighted in bold)

of an organization, and therefore, the implications of ERP implementation could be fundamentally wider than of any traditional transaction, or functional, system" [21]. Therefore, we use this sub-class of EAM-enabled transformation to empirically evaluate and discuss parts the research model proposed above. In figure 2, the according parts of the research model used in the following evaluation and in the discussion are highlighted in bold.

5   Empirical Evaluation

As a proof of concept, we analyze ERP implementation projects as an exemplary sub-class of EAM-enabled transformation situations.

5.1   Outline

ERP systems are integrated software solutions used to manage any organization's resources [22]. In view of the sums invested by organizations, the success rate of ERP implementation projects is not high [23]. In the literature, various critical success factors and issues in ERP implementation projects have been identified. Often, it has been indicated that a combination of inadequate preparedness and inappropriate project management have been responsible for the low-success rate of ERP implementations [23]. Al-Fawaz et al. [24] identified top management support, having an appropriate business plan and vision, re-engineering business process, effective project management, user involvement, and education and training as critical success factors in ERP implementation projects. According to Bradley [25], choosing the right full time project manager, training of personnel, and the presence of a champion relate to project success. The use of consultants, the role of management in reducing user resistance and the use of a steering committee to control the project do not appear to differentiate successful and unsuccessful projects [25]. Integration of ERP planning with business planning, reporting level of the project manager, and active participation of the CEO beyond project approvals, resource allocation and occasional project review, are not found to be critical factors of success [25].

   As this roundup shows, the factors promoting and hindering an ERP implementation project's success are manifold. Therefore, in order to derive the major leverages for EAM in ERP implementation projects, we identify the most pressing issues in

ERP implementation projects on the basis of mean added project duration and probability of occurrence of each issue. By addressing these concrete issues, EAM can be used in an efficient and effective manner. In the discussion (cf. section 6), we will map the findings of this section to the EAM approaches.

5.2 Survey Design and Data Collection

The data was collected using the Delphi method. The Delphi method is a method for structuring, e.g., the gathering of current and historical data not accurately known or available, on the basis of a group communication process allowing a group of individuals, as a whole, to deal with a complex problem [26]. For our research, the Delphi method seems especially appropriate, as it uses the combined wisdom of a focus group to estimate occurrences and the evolution of trends (ongoing phenomena) when there is no source of factual data and a basis for opinion exists [27]. A written questionnaire was used to structure the interviews. Specialists and executives have been invited who actively are engaged in large-scale, IT enabled transformation projects.

For the design of the survey, we roughly adhered to the process as proposed by [28]. The survey instrument consisted of 27 items. For each of the items, the mean added project duration in relation to the planned project duration and the probability of the occurrence of project duration enhancement were measured on a percentage scale. The final survey included items as depicted in table 5.

5.3 Data Analysis

By means of an exploratory analysis we investigate the status quo of transformation projects in practice. In order to elucidate the predominant design situations, data is examined by cluster analysis.

The purpose of clustering, a form of combinatorial data analysis, is to investigate "a set of objects in order to establish whether or not they fall […] into groups […] of objects with the property that objects in the same group are similar to one another and different from objects in other groups." [29] At the beginning of the investigation, these groups are unknown and need to be determined. Various clustering methods exist. They can be categorized by the type of algorithm used to obtain the clusters.

Agglomerative, divisive, incremental, direct optimization, and parallel algorithms can be distinguished [29]. Agglomerative algorithms start with n clusters, each containing a single object. One by one at each step, the number of clusters is reduced. According to [30], agglomerative algorithms have the largest significance in practice. Therefore, we utilize a clustering method which is based on an agglomerative algorithm. Concerning the selection of the clustering method it should be noted "that there is usually no uniquely obvious method of analyzing the data", as it is highly dependent on the selection of an appropriate clustering criterion [29].

The Ward method is used for clustering, as it finds very good partitions and reveals the appropriate number of clusters with a similar number of observations in each cluster at the same time [31, 32]. Ward's method starts with individual cases each forming a separate cluster and progresses by combining them into clusters until each and every case is in the same cluster. The decision which clusters to merge next is based on minimizing the sum of the squared Euclidean distance of each case from the

Table 5. Survey items

Description	Mean added project duration in %	Probability of occurrence in %
Poor interpersonal or functional skills of key people	24.3	26.4
Poor interpersonal or functional skills of project team	24.6	24.3
Poor (people) change management (e.g. insufficient comm. and user involvement)	21.9	35.0
Poor process management (e.g. no process owners defined, no policies for standardised processes in place)	23.4	25.7
Heavy ERP modification to adapt the IT solution to the customer's processes	31.8	29.1
Overly extensive business blueprint and system documentation (e.g. functional specification)	13.1	17.3
Incomplete specification of functional requirements	19.1	44.3
Lack of stakeholder commitment (e.g. resistance of business managers to "IT-project")	26.1	37.5
Insufficient scope management (e.g. unexpected changes to project scope)	19.6	34.5
Project team is geographically separated	8.2	51.6
Large number of part-time project team members	7.1	38.5
High fluctuation within project team	13.6	25.0
Insufficient planning and control of the project portfolio	17.1	26.6
Overestimation of effort	12.4	12.6
Insufficient monitoring of project progress and efforts spent	13.6	19.7
Inadequate overall project approach (e.g. starting from scratch instead of standard best practice process and configuration)	30.8	22.9
Insufficient available standard best practice configuration	18.7	23.9
Insufficient available documentation of best practice standard (process, technical specification, etc.)	16.2	31.1
Insufficient/inadequate pre-existing test cases	9.6	30.6
Insufficient existing demo system	5.4	27.0
Insufficient existing methodology to implement customer-specific requirements	13.8	21.2
Insufficient tools to implement existing methodology (e.g., automated configuration, testing)	7.6	20.2
Insufficient material/tools to enable organisational change	10.4	28.2
Unclear definition of roles between customer, vendor, and partner	8.0	22.1
Insufficient alignment or coordination between vendor(s) and system integrator(s)	9.0	19.6
High complexity of integration between ERP system components	6.8	11.8
High complexity of integration between ERP system and third-party components	17.7	31.4

mean of its cluster. Other clustering algorithms and distance measures were also tested, but the Ward method in combination with the squared Euclidean distance produced the best results in terms of interpretability, context, and purpose of the study at hand [33]. The results of the clustering algorithm are graphically represented in form of the design situations in figure 3.

Fig. 3. Design situations in transformation projects

6 Discussion

The upper right quadrant in figure 3 contains the most pressing issues in relation to mean added project duration and probability of occurrence in ERP implementation projects.

In order to address these issues, EAM should firstly support work scope control. This can be achieved by documenting and checking the completeness of functional requirements specifications-not only during the initial project stages, but on a regular basis to keep track of changes and scope adjustments. Furthermore, EAM should document where and why functional requirements divert significantly from ERP functionalities, thereby inducing potentials for costly ERP modifications.

Secondly, EAM should support the creation and maintenance of analyses and visualizations that are targeted at important business stakeholders' needs. ERP projects need not only to be managed, but also be communicated and visualized as business projects. As a consequence, business stakeholders need to be involved in architectural governance bodies.

Thirdly, EAM must incorporate not only the new, well specified ERP architecture components, but also the relevant legacy and third-party portions of the application landscape that create integration requirements. Although creating additional project efforts, fragmented and partial EA support will not be capable to address large-scale ERP integration issues.

Finally, best practice standards for processes, technical specifications, etc. need to be incorporated in EAM representations and analyses. This means that in addition to as-is and to-be EA artefacts, also reference artefacts from different sources and with different scopes need to be a part of the corporate EAM repository.

The four major leverages for EAM in ERP implementation projects are summarized in table 6.

Table 6. Major leverages for EAM in ERP implementation projects

#	Name	Description
1	Control of work scope	Specification of functional requirements and scope management
2	Stakeholders	Stakeholder communication and commitment
3	Legacy	Complexity of integration between ERP system and third-party components and degree of ERP modification to adapt the IT solution to the customer's processes
4	Methodology	Documentation of best practice standards

In the following, we discuss which EAM approach is the most promising one in order to address certain issues in ERP implementation projects.

The control of the work scope is a very important task in organizational transformation [34]. Levene and Braganza suggest to focus on the interfaces between different stakeholders to support the changing shape of transformation projects and to support the inter-stakeholder communication [34]. As regards the EAM approach, the "Mature EAM" approach with its holistic EA understanding and a progressive perception of EA within the business and management units seems especially appropriate for this. Furthermore, the "Mature EAM" approach and also the "Initial EAM" approach provide the necessary methodology, e.g. advanced architectural design paradigms, for technology enabled growth – a characteristic missing in the "IT-biased EAM" approach with its rather limited architectural understanding. Both the "IT-biased EAM" approach and the "Initial EAM" approach provide means, e.g. tools and models, to address issues because of high complexity imposed by legacy systems. To some degree, this is also true for the "Mature EAM" approach, but this approach goes beyond addressing complexity because of legacy systems and tackles the topic of technology enabled growth in a holistic manner by also including stakeholders, methodological aspects, and the changing nature of such transformation projects. Figure 4 summarizes this mapping of EAM approaches to the major leverages in ERP implementation projects as a sub-class of the "Technology enabled growth" transformation situation.

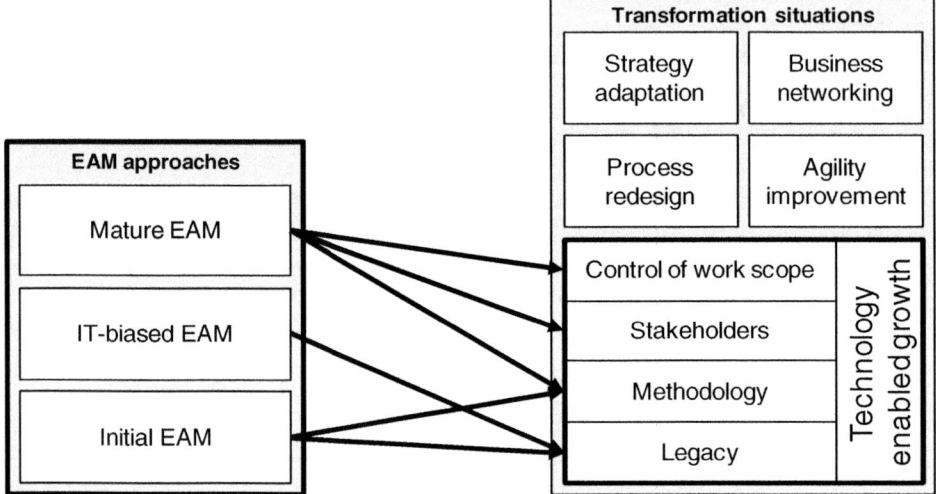

Fig. 4. Mapping EAM approaches to the major leverages in the "Technology enabled growth" transformation situation.

7 Conclusion

In this paper, we gave an overview of contingency theory, situational EAM and situational transformation management. Based on this, we developed a research model that integrates these topics and suggests that the success of certain transformation projects depends on the design factors contingent to the specific situation of an organization. On the basis of an empirical study on ERP implementation projects, we identified the most pressing issues in ERP implementation projects as regards mean added project duration and probability of occurrence of each issue and clustered these issues in the form of four major leverages for EAM in ERP implementation projects. We concluded that a certain EAM approach is most promising for a certain transformation situation. If validated, the findings enable the construction of situational, and therefore more effective, artefacts like e.g. method fragments, reference models, design principles, and ultimately methods for supporting respective design problems in organizations.

With the empirical evaluation, we only covered transformation situations of the "Technology enabled growth"-type. In future research, this should be extended to cover the other four transformation situations as identified by Baumöl [3]. Furthermore, the discussion only provides an argumentative mapping of the design situations in transformation projects to the EAM approaches and transformation situation. Future research should provide empirical support for the outlined relationships.

Acknowledgement

The empirical evaluation of this research has been conducted as a part of SAP's Business Transformation Academy initiative in which the University of St. Gallen is involved. This research is a joint deliverable of the initiative and we acknowledge SAP's effort in developing and distributing the questionnaires, securing a good response rate and interpreting the survey results.

References

1. Winter, R.: Problem Analysis for Situational Artefact Construction in Information Systems (to appear 2010)
2. Aier, S., Riege, C., Winter, R.: Classification of Enterprise Architecture Scenarios – An Exploratory Analysis. Enterprise Modelling and Information Systems Architectures 3(1), 14–23 (2008)
3. Baumöl, U.: Strategic Agility through Situational Method Construction. In: Proceedings of (2005)
4. Hevner, A.R., March, S.T., Park, J., Ram, S.: Design Science in Information Systems Research. MIS Quarterly 28(1), 75–105 (2004)
5. March, S.T., Smith, G.F.: Design and Natural Science Research on Information Technology. Decision Support Systems 15(4), 251–266 (1995)
6. Baskerville, R.L., Pries-Heje, J., Venable, J.: Soft design science methodology. In: Proceedings of Proceedings of the 4th International Conference on Design Science Research in Information Systems and Technology (DESRIST 2009), pp. 1–11. ACM, New York (2009)
7. Becker, J., Delfmann, P., Knackstedt, R.: Adaptive Reference Modeling: Integrating Configurative and Generic Adaptation Techniques for Information Models. In: Becker, J., Delfmann, P. (eds.) Reference Modeling, pp. 27–58. Physica, Heidelberg (2007)
8. Becker, J., Delfmann, P., Knackstedt, R., Kuropka, D.: Konfigurative Referenzmodellierung. In: Becker, J., Knackstedt, R. (eds.) Wissensmanagement mit Referenzmodellen. Konzepte für die Anwendungssystem- und Organisationsgestaltung, pp. 25–144. Physica, Heidelberg (2002)
9. Bucher, T., Klesse, M., Kurpjuweit, S., Winter, R.: Situational Method Engineering - On the Differentiation of "Context" and "Project Type". In: Proceedings of Situational Method Engineering - Fundamentals and Experiences, Boston, pp. 33–48. Springer, Heidelberg (2007)
10. Fiedler, F.E.: A Contingency Model of Leadership Effectiveness. Advances in Experimental Social Psychology 1, 149–190 (1964)
11. Donaldson, L.: The Contingency Theory of Organizations. Sage, Thousand Oaks (2001)
12. Graubner, M.: Task, firm size, and organizational structure in management consulting. In: An empirical analysis from a contingency perspective. DUV, Wiesbaden (2006)
13. Miles, R.E., Snow, C.C., Meyer, A.D., Coleman Jr, H.J.: Organizational Strategy, Structure, and Process. Academy of Management Review 3(3), 546–562 (1978)
14. Venkatraman, N., Camillus, J.C.: Exploring the Concept of "Fit" in Strategic Management. The Academy of Management Review 9(3), 513–525 (1984)
15. White, R.E.: Generic Business Strategies, Organizational Context and Performance: An Empirical Investigation. Strategic Management Journal 7(3), 217–231 (1986)

16. Aier, S., Gleichauf, B.: Application of Enterprise Models for Engineering Enterprise Transformation. Enterprise Modelling And Information Systems Architectures 5(1), 56–72 (2010)
17. Winter, R., Bucher, T., Fischer, R., Kurpjuweit, S.: Analysis and Application Scenarios of Enterprise Architecture - An Exploratory Study. Journal of Enterprise Architecture 3(3), 33–43 (2007)
18. Ylimäki, T., Halttunen, V.: Method engineering in practice: A case of applying the Zachman framework in the context of small enterprise architecture oriented projects. Information Knowledge Systems Management 5(3), 189–209 (2006)
19. Baron, J.N., Hannan, M.T.: Organizational Blueprints for Success in High-Tech Start-Ups: Lessons from the Stanford Project on Emerging Companies. California Management Review 44(3), 8–36 (2002)
20. Moreton, R.: Transforming the organization: the contribution of the information systems function. Journal of Strategic Information Systems 4(2), 149–163 (1995)
21. Nandhakumar, J., Rossi, M., Talvinen, J.: The dynamics of contextual forces of ERP implementation. Journal of Strategic Information Systems 14(2), 221–242 (2005)
22. Basoglu, N., Daim, T., Kerimoglu, O.: Organizational adoption of enterprise resource planning systems: A conceptual framework. Journal of High Technology Management Research 18(1), 73–97 (2007)
23. Carton, F., Adam, F., Sammon, D.: Project management: a case study of a successful ERP implementation. International Journal of Managing Projects in Business 1(1), 106–124 (2008)
24. Al-Fawaz, K., Al-Salti, Z., Eldabi, T.: Critical Success Factors in ERP Implementation: A Review. In: Proceedings of, pp. 1–9 (2008)
25. Bradley, J.: Management based critical success factors in the implementation of Enterprise Resource Planning systems. International Journal of Accounting Information Systems 9(3), 175–200 (2008)
26. Linstone, H.A., Turoff, M. (eds.): The Delphi Method: Techniques and Applications (2002)
27. Gray, P., Hovav, A.: From Hindsight to Foresight: Applying Futures Research Techniques in Information Systems. Communications of the Association for Information Systems 22(12), 211–234 (2008)
28. Moore, G.C., Benbasat, I.: Development of an Instrument to Measure the Perceptions of Adopting an Information Technology Innovation. Information Systems Research 2(3), 192–22 (1991)
29. Gordon, A.D.: Hierarchical Classification in Phipps Arabie. In: Hubert, L.J., De Soete, G. (eds.) Clustering and Classification, pp. 65–121. World Scientific Publishing, River Edge (1996)
30. Härdle, W., Simar, L.: Applied Multivariate Statistical Analysis. Springer, Berlin (2003)
31. Ward Jr., J.H.: Hierarchical Grouping to Optimize an Objective Function. Journal of the American Statistical Association 58(301), 236–244 (1963)
32. Hair Jr, J.F., Black, B., Babin, B.: Multivariate Data Analysis, 6th edn. Prentice Hall, Australia (2006)
33. Han, J., Kamber, M.: Data Mining - Concepts and Techniques, 2nd edn. Morgan Kaufmann, San Francisco (2006)
34. Levene, R.J., Braganza, A.: Controlling the work scope in organisational transformation: a programme management approach. International Journal of Project Management 14(6), 331–339 (1996)

Building Blocks for Enterprise Architecture Management Solutions

Sabine Buckl, Thomas Dierl, Florian Matthes, and Christian M. Schweda

Technische Universität München, Institute for Informatics 19,
Boltzmannstr. 3, 85748 Garching, Germany
{sabine.buckl,dierl,matthes,schweda}@in.tum.de
http://www.systemcartography.info

Abstract. Enterprise architecture (EA) management has become a commonly accepted means to guide enterprises in transformations responding to their ever changing environment. Organizations seeking to establish an integrated and effective EA management function are typically faced with a challenging lack of standardization in the field. Although the topic is heavily researched by practitioners, researchers, standardization bodies, and tool vendors, no commonly accepted understanding of the scope, reach, and focus of EA management exists. This fact can be explained by the distinct organizational structures, contexts, cultures, and requirements, which are specific for each enterprise and therefore ask for an enterprise-specific realization of the EA management function.

In response to the aforementioned challenge this article presents *building blocks for EA management solutions* (BEAMS). BEAMS on the one hand provides practical guidance for organizations to support the design and development of an organization-specific EA management function by presenting method and language building blocks, which can be selected and configured based on the specificities of the organization under consideration, i.e. the organizational context and the goals pursued. On the other hand BEAMS gives hints for researchers willing to contribute to the discipline of EA management. The theoretic discussion on the developing BEAMS approach is complemented by an example to illustrate the applicability of the approach. Finally, a critical reflection of the achieved results is given and future areas of research are discussed.

Keywords: EA management function, building blocks, patterns, situational method engineering, design theory nexus.

1 Introduction

In a rapidly changing economic, technical, and regulatory environment, the flexibility to adapt to changes as well as the ability to implement new business capabilities quickly are both vitally important for companies regardless of their type and size. Emerging paradigms as service oriented architectures (SOA), domain-specific languages or model driven development claim to be helpful in this context, but when it comes to their implementation in an organization, subtle difficulties arise. This can be exemplified with the implementation of an SOA, but

F. Harmsen et al. (Eds.): PRET 2010, LNBIP 69, pp. 17–46, 2010.

holds for other paradigms as well: restructuring the IT landscape of an organization towards services is a long-lasting endeavor, needing not only quite a few information on the applications, but also on the connected business processes and business objects. Most likely such information is not present at the beginning of an SOA transformation program, but has to be gathered in an extensive process. Even if the information is available, the transformation program cannot assume that 'the world keeps from turning', i.e., the organization does not stop changing. Therefore, the transformation program has to be continuously realigned with the change and maintenance projects that are executed in parallel. In this respect, realizing all benefits of an SOA transformation is only possible in an environment, where business and IT *development* are aligned.

This mutual alignment goes beyond a mere *provider role* of the IT, in which IT resorts itself to solely fulfilling business requirements. IT in contrast has to take an *enabler role*, proactively seeking to increase flexibility and adaptability to foster the agility of the overall organization. This two-fold role of the IT well illustrates the very core of business-IT alignment (cf. [23,35,52]), which could have also been described conversely from a business perspective. In consequence, mutual alignment is a goal best to be approached from both perspectives – a business and an IT perspective – and is hence not in the focus of the management functions for business or IT management, respectively. This calls for a management function with an embracing management subject spanning business- and IT-related concepts, but most preferably also accounting for *crosscutting aspects*, as strategies and projects. The latter is especially necessary as a managed evolution of the organization inevitably connects to the strategies as drivers of organizational change and the projects as its vehicles. This holistic understanding of the organization actually is the one incorporated in enterprise architecture (EA), i.e. the architecture of the enterprise which in accordance with the ISO Standard 42010 [26] can be defined as follows:

> Enterprise architecture (EA) is the fundamental conception of the enterprise in its environment, embodied in its elements, their relationships to each other and to its environment, and the principles guiding its design and evolution.

The management of the EA forms a management discipline that seeks to address the aforementioned topic of mutual alignment by taking the embracing perspective of the overall EA. This new management discipline has – not surprisingly – attracted practitioners and researchers seeking for guidance on how to conduct and perform EA management. Research in this area is typically conducted in close cooperation and interaction with an organization willing to practically apply the research results. On the one hand this opens the door for "developing case studies" (cf. van Aken in [49, page 232]) by employing an intrinsically motivated industry partner. On the other hand, industry-funded research projects usually underly the partnering organization's pace and hence often force early delivery of results, which aggravates the development of comprehensive theoretical underpinnings. Thus, researchers in the area of EA management are challenged to ensure that their research conducted in close interaction with organizations does not degenerate into "routine design" that according to Hevner et al. in [24, page 82] must be distinguished from design science. In contrast, the close cooperation can be used to contribute to theory building e.g. via extracting case studies (cf. van Aken in [49] as well as Eisenhardt and Graebner in [15]). Building on a figure from Gehlert et al. in [20, page 442] that illustrates the twofold relation between theory and design according to Nunamaker et al. in [40], we highlight the interplay between design and theory building (cf. Figure 1).

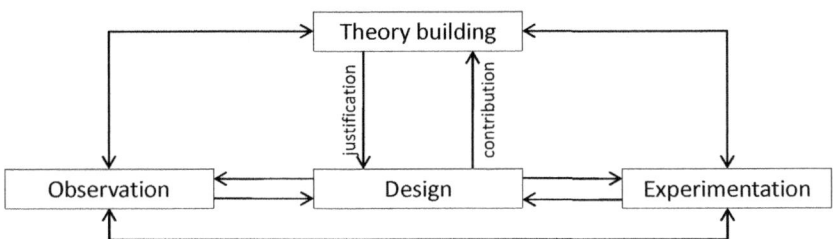

Fig. 1. The interplay of design and theory building [20,40]

In the light of this interplay and against the backing environment in which EA management research is typically carried out, the first research question guiding our subsequent considerations can be derived.

How can researchers on the one hand contribute to the knowledge base of EA management, and at the same time deliver early results applicable in practice?

An interesting area for contributing to the knowledge base of EA management is concerned with the structure of the EA management function itself, as currently, no commonly accepted step-by-step guidelines for performing EA management exist. This absence might be caused by the fact, that no EA management process model detailing the management function has yet gained prominence. Some researchers even doubt the existence of a *one-size-fits-them-all* approach, but expect the management function to be organization-specific (cf. [5,31,50,48]). This situation is similar to the one in software development, where albeit a general agreement on important activities as e.g. *requirements elicitation* or *testing*, various process models exist, which strongly differ concerning the linkages between the different activities and the level of detail in which the different activities are described[1]. The situation of EA management is even more complicated than the one in software development. The goals of a software development process are typically agreed upon as "developing a software system in time, with the required functionality and quality, as well as within the planned budget" [47]. The objectives of an EA management initiative in contrast vary widely. While typical EA management goals can be summarized on an abstract level, they have to be substantiated during the establishment of an appropriate organization-specific

---

[1] For a in-depth discussion of different software development process models see [37].

management function in order to identify the elements of the EA relevant for the initiative. Reducing maintenance costs via standardization can for instance be performed on different levels, e.g. on business processes, business support provided by business applications, or on a more technical infrastructure level.

Besides the variety of different goals, which need to be appropriately addressed by the EA management function, the organizational context, in which the function has to be embedded and operated, influences the suitability of an EA management approach. While in a smaller company with a familiar atmosphere, the simple communication of architectural principles might be sufficient to ensure project compliance, a more hierarchical corporate culture might demand for the establishment of quality gates, e.g. architecture reviews prior to the project start as well as controls after realization of the project to ensure adherence to architectural principles and standards. Some of the existing approaches even stress the fact that they have to be adapted to the context of the applying organization (cf. "adapting the ADM [architecture development method]" in [48, page 56 seq]) but typically abstain from providing information on how to perform these adaptations. A better situation can be identified regarding the goals pursued by the different approaches, which are typically detailed on an abstract level as mentioned before. This leads us to the second research question of this article

> How does a configurable approach to design EA management functions look alike?

In this article, we answer the aforementioned research questions by presenting *building blocks for EA management solutions* (BEAMS). The BEAMS approach is based on a conceptualization of Pries-Heje and Baskerville, who introduced the concept of a design theory nexus in [41], the prefabrics of the pattern-based approach to EA management presented by Buckl et al. in [6] and Ernst in [17], and the situational method engineering as discussed by Harmsen in [22]. The resulting approach is presented in Section 3 and complemented with a constructed case providing an example of how BEAMS can be applied (cf. Section 4). With BEAMS being a relatively new approach, currently a comprehensive practical evaluation has yet not been conducted. To provide indications on the suitability and applicability of the approach, a comparison with prominent approaches representing the state-of-the-art of EA management practice is given in Section 5. Final Section 6 provides a critical reflection of the achieved results and hints to further areas of research.

2   Prefabrics from Related Disciplines

Developing an EA management function, which suits the specific needs of the using organization is a challenging task. Taking into account the rich literature on the subject EA management as presented by Langenberg and Wegman in [32] as well as Schönherr in [45], the development activity can be understood as task in which different competing solutions offered by existing EA management approaches are compared and evaluated in respect to their suitability. The evaluation a) is thereby based on the goals pursued by the EA management initiative and b) accounts for the organizational context to embed the EA management function into. Two different approaches to perform such evaluation and to facilitate the aforementioned challenge are discussed below. First, the construct of a design theory nexus (DTN), which provides "a set of constructs and methods that enable the construction of models that connect numerous design theories with alternative solutions" [41, page 733], is introduced. Secondly, the idea of situational method engineering is presented, which describes how a method can be "tailored and tuned to a particular situation" (cf. Harmsen in [22, page 25]).

2.1   A Design Theory Nexus for Competing Solutions

In [41], Pries-Heje and Baskerville present the idea of a DTN as means to connect existing competing solutions, i.e. design theories, for a problem domain. A DTN is no simple framework connecting these solutions, but further helps "decision makers in choosing which of the theories are most suitable for their particular goals and their particular setting" [41, page 733], i.e. the organizational environment as well as the goals to pursue. Pries-Heje and Baskerville discuss that their approach is useful in cases of solving *wicked problems*. Establishing an EA management function forms a wicked problem, for which a plethora of competing solutions exist. A DTN instantiation for EA management can be developed, which provides assistance in choosing a suitable EA management approach according to the organizations' goals and organizational context. According to Pries-Heje and Baskerville in [41, page 743], a DTN instantiation consists of the following four constructs depicted in Figure 2:

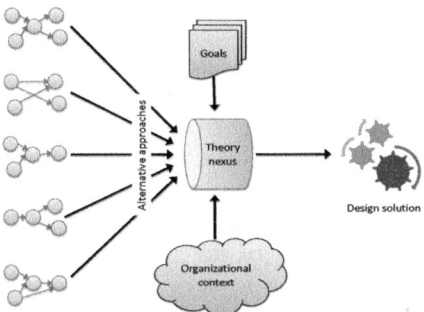

Fig. 2. Components of a DTN according to Pries-Heje and Baskerville in [41]

- **Goals** describe what the design solution is intended for.
- **Organizational context** refers to contingencies outside of the people involved.
- **Design theory nexus** defines the connection point at which the competing theories are bound with realities into a design solution.
- **Design solution** represents the result constructed from dissimilar decision alternatives.

The instantiation, i.e. construction, of a DTN according to Pries-Heje and Baskerville (cf. [41]) follows a five step approach. In the first step, the available approaches in the area under consideration are examined, e.g. via a literature analysis. In a second step, the identified competing theories are investigated for explicit or implicit conditions that must hold for the approach to achieve the highest utility. Here, it has to be noted that these conditions might not match, i.e. be *asymetric*, for any pairing of the theories. The third step assesses the identified conditions for practical relevance and formulates them to assertions. In the fourth step, a decision-making process for evaluating the assertions is undertaken. Final step five combines the approaches, conditions, assertions, and the process into a tool, which supports the evaluation of the fit for each approach in a given situation.

2.2   Situational Method Engineering

Motivated by the plurality of proposed methods for information system engineering as well as the increasing application area diversification and complexity, Harmsen presented in [22] an approach to *situational method engineering*. The driving idea behind situational method engineering can be summarized by the following quote: "There is no method that fits all situations" [22, page 6]. Introducing the term *controlled flexibility* Harmsen elicits requirements for a method engineering approach, which accomplishes method standardization and at the same time is flexible enough to match the situation at hand. A *situation* thereby refers to the combination of circumstances at a given point in time in a given organization [22]. In order to address these requirements, for each situation a suitable method – so-called situational method – is *constructed* that takes into account the circumstances applicable in the respective situation. In the construction process uniform method fragments are selected, which can be configured and adapted with the help of formally defined guidelines.

The generic process to constructing situational methods consists of four steps. Input to the configuration process is the specific situation in which the method should be applied, e.g. the environment of the initiative, including users, organizational culture, management commitment, etc. This situation is analyzed in the first step (*characterization of the situation*) to describe the application characteristics. This information is used in the second step (*selection of method fragments*) to select suitable method fragments from the method base. Heuristics can thereby be applied to foster the selection process. In the third step (*method assembly*) the method fragments corresponding with the situation characterization are combined to a situational method. During assembling method fragments, aspects like completeness, consistency, efficiency, soundness, and applicability are accounted for (cf. [22]). The actual use of the constructed situational method is performed in the last step (*project performance*). Figure 3 gives an overview on the construction process and the relationships between the different steps.

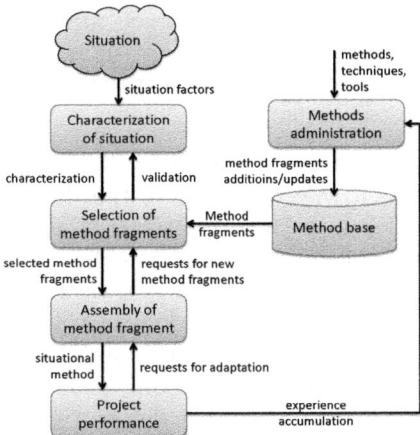

Fig. 3. The process of situational method engineering according to [22]

Complementing the construction process of a method situated for a given environment, Harmsen introduces in [22] the activity *methods administration* that captures methodical knowledge, i.e. adds or updates method fragments, if necessary, based on feedback from the project performance step.

Developing a harmonization in the area of method engineering and at the same time emphasizing on the influence of the particular situation a method should be applied in, represents the core idea in situational method engineering as presented by Harmsen in [22]. The state of IS engineering described by Harmsen is quite similar to the one in developing and designing an organization-specific EA management function. A multitude of approaches exists but none of these has gained prominence due to the situation- or organization-specificity of the subject. Therefore, we propose an approach that picks up the idea of the DTN presented by Pries-Heje and Baskerville in [41] and the approach of situational method engineering presented by Harmsen in [22].

2.3   A Pattern-Based Approach to EA Management

Patterns have a long history as useful means for documenting re-usable solutions for recurring problems in a complex domain, dating back to Alexander [3], who introduced patterns as "coherent and modular solutions to specific problems". Further publications (cf. Buschmann et al. [12] or Gamma et al. [19]) have refined the term pattern and put forward structuring guidelines for the description of patterns. A broadly accepted structure is presented by Buschmann et al. in [12], according to which a pattern is constituted of the following elements:

- **Context description**, which is concerned with causes and environmental factors that may have lead to the problem that the pattern solves.
- **Problem description**, which alludes to the issues and difficulties that occur in many contexts and may be solved with the pattern. Thereby, the description expatiates on conflicting forces that comprise the problem.

- **Solution description**, which explains the steps to be taken and the concepts to be used in order to solve the corresponding problem.
- **Consequence description**, which refers to consequences that may be caused by applying the pattern to the given problem.

Building on the idea of patterns, Buckl et al. [6] coined the term of the "EA management pattern" as a way to structure the domain of EA management. In [16], Ernst further develops this idea towards a pattern language for EA management. This pattern language introduces three types of patterns, namely *methodology pattern (M-Pattern)*[2], *viewpoint pattern (V-Pattern)*, and *information model pattern (I-Pattern)* that are used to develop an organization-specific EA management function. These three types of *EA management patterns* describe constituents of proven-practice solutions for EA management as found in literature but also in practice (cf. [13]). The different types of patterns contribute different parts to an EA management function, detailed as follows:

- **M-Patterns** describe management methods (and processes) that solve a specific EA-related problem. Thereby, a pattern provides step-by-step guidance and information on what and how to do.
- **V-Patterns** describe viewpoints, i.e., types of visualizations that are employed by an M-Pattern in order to communicate solution-relevant information about the EA.
- **I-Patterns** describe conceptual models, whose concepts are instantiated to documentations of solution-relevant parts of the overall EA.

Buckl et al. describe in [6] how the three types of EA management pattern can together be used to design an organization- and problem-specific EA management function. The context descriptions provided by the patterns are explored during this phase in order to select the appropriate patterns that optimally fit the organizational context. The problem descriptions are the starting points for selecting the "right" EAM pattern, i.e. those patterns that solve the organization-specific problems. If different patterns were applicable to similar problems and hence were to be decided upon during the design process, the context and consequence description could provide additional help to choose the patterns that optimally balance desired outcomes and side-effects. Finally, the interrelationships between the patterns, which are described as part of the pattern language, support the identification of patterns that might also apply in the given context or may be helpful for solving related problems. After having selected the appropriate patterns, the methods, viewpoints, and conceptual models described therein have to be integrated into a management function. This final design step requires method engineering capabilities, as especially the M-pattern as described in [13] do not provide integration artifacts. The same is true for the V- and I-Pattern, which have to be integrated into a comprehensive EA modeling language. While some issues on integration are discussed by Buckl et al. in [6], dedicated integration artifacts and mechanisms are not provided.

---

[2] Being more clear with respect to the terminology, these patterns should be alluded to as "method patterns".

The absence of integration related prescriptions may be explained with the focus and the nature of the approach. Patterns are solutions observed in practice, i.e. describe real-world solutions, and are not engineered or developed towards an integrated knowledge base. This can be exemplified with the M-patterns that describe methods and processes for conducting EA management, but are not concerned with designing an EA management function. A design method for EA management function would have to provide additional guidance for

- **selecting** the appropriate EA management patterns, especially in case different of them are applicable,
- **integrating M-patterns** into a consistent EA management process, especially avoiding process redundancies, and
- **integrating V- and I-patterns** into an EA modeling language, especially accounting for the information demands of the viewpoints.

The subsequently presented approach refines the EA management patterns described in [13] to address the aforementioned issues of integration. The patterns are reorganized and rewritten to redundancy free and composeable building blocks for EA management solutions.

3 BEAMS – Components and Design

The BEAMS approach presents a DTN instantiation for situational EA management based on the groundworks presented in Section 2. In Section 3.1 the structure and interplay of the components of BEAMS is described, reflecting the core dichotomy of EA management – method and language – as manifested in the EAM pattern approach (cf. Section 2.3) and discussed by Schelp and Winter in [44]. Based on the common understanding of the constituents, we discuss the construction process of BEAMS in Section 3.2 utilizing the five-step method as proposed by Pries-Heje and Baskerville in [41].

Prior to presenting BEAMS and its components, a central design principle of the approach should be introduced, namely the principle of "loose coupling" between the building-blocks. By this term, we describe a characteristic of the inner organization of BEAMS. As opposed to a pattern-language the building-blocks are not interlinked by *explicit*, i.e. *material*, relationships. The actual relationships are of *implicit* nature, i.e. constitute *formal* relationships, that may be derived from the building-blocks relations to the underlying framework and stratified terminology. Put in other words, the BEAMS' underlying framework supplies an ontology, on which the building-blocks are built in a way that their (formal) interrelations may be derived from linkages to the same framework constituents. As a consequence, this design principle helps to develop BEAMS (cf. Section 3.2) from manifold sources without having to intermesh the different design theories and prescriptions in a dense web of newly established relationships.

3.1 Components of BEAMS

The idea of interrelating competing solutions to design an organization-specific EA management function can only be realized against the basis of a common understanding and terminology. Multiple publications have targeted this topic. Schönherr showed in [45] that the discipline of EA management is not yet developed a fully consistent terminology, but is on the way to do so. In a similar vein, Schelp and Winter discuss in [44] that different "language communities" in EA management research exist, although certain degree of convergence has recently been reached. Nevertheless, when it comes to distinct aspects of EA management multiple approaches agree on a common understanding regardless of some terminological differences. One of these aspects is the question of the fundamental activities and tasks of EA management. Buckl et al. revisit in [11] the different perspectives on this topic as put forward in literature taking into account the pattern approach (cf. [13]) as well as the approaches of Frank [18], Wegmann [51], Hafner and Winter [21], Niemann [38], Schekkerman [43] and The Open Group [48]. Based on the literature, Buckl et al. devise a method framework for EA management consisting of four activities as shown in Figure 4:

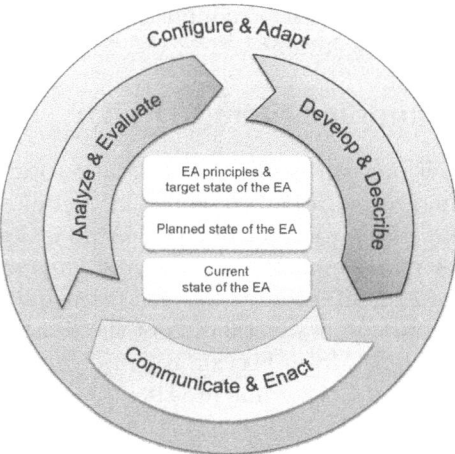

Fig. 4. Method framework of BEAMS

Develop & describe a state of the EA, either a current state describing the as-is architecture, a planned state or a target state, i.e. an EA vision.

Communicate & enact architecture states and principles to EA-relevant projects and to related management functions, as project portfolio management.

Analyze & evaluate architectural scenarios (planned states) or analyze whether a planned state helps to achieve the target state or not.

Configure & adapt the EA management function itself, i.e. decide on the management concerns, goals, and methods.

Against the background of the method framework reverberating through related EA management literature, we can devise the core structure of BEAMS in context as shown in Figure 5. The core constituents of BEAMS as a DTN instantiation for the field of EA management design are:

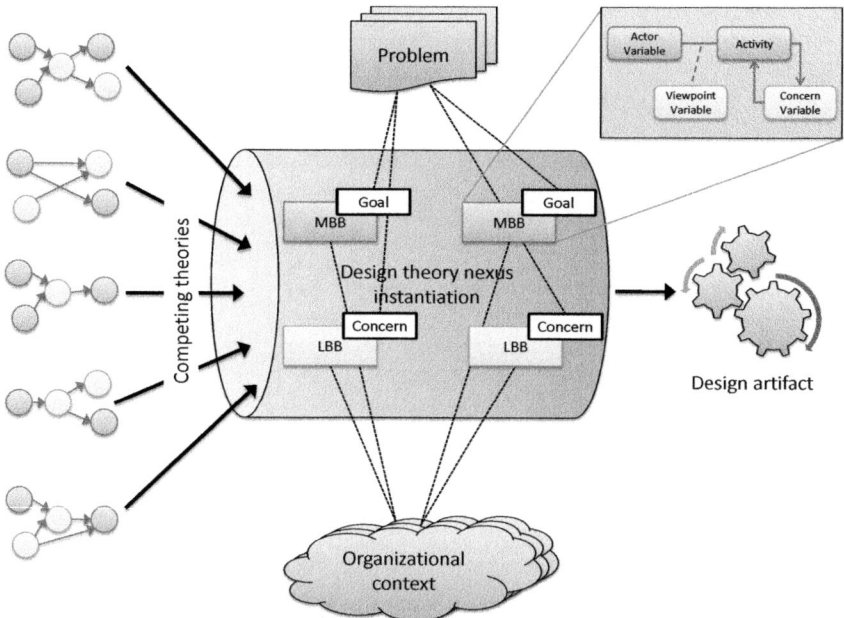

Fig. 5. Components of BEAMS

Competing theories: The competing theories represent the knowledge base from which BEAMS is built. Reflecting the nature of EA management as a practice-oriented field of research, BEAMS builds on solutions with have been proven valuable in practice, e.g. patterns, best practices, case studies.

Problem: A problem represents the issue to be solved by applying the theory. A problem in the area of EA management typically consists of a

goal representing an abstract objective, e.g. increase homogeneity, provide transparency, and a

concern, i.e. area of interest in the enterprise, e.g. business support, application systems.

Organizational context: The organizational context represents the situation in which the EA management function operates. Typical factors which are considered in the organizational context are the organizational culture, management commitment, stakeholders, etc.

Building block (BB): The building blocks form the solution models to be combined to an organization-specific EA management function. Reflecting the dichotomy of method and language, two kinds of building blocks exist,

method building block (MBB) describing who has to perform which tasks in order to address a problem in the situated context and

language building block (LBB) referring to which EA-related information is necessary to perform the tasks and how it can be visualized.

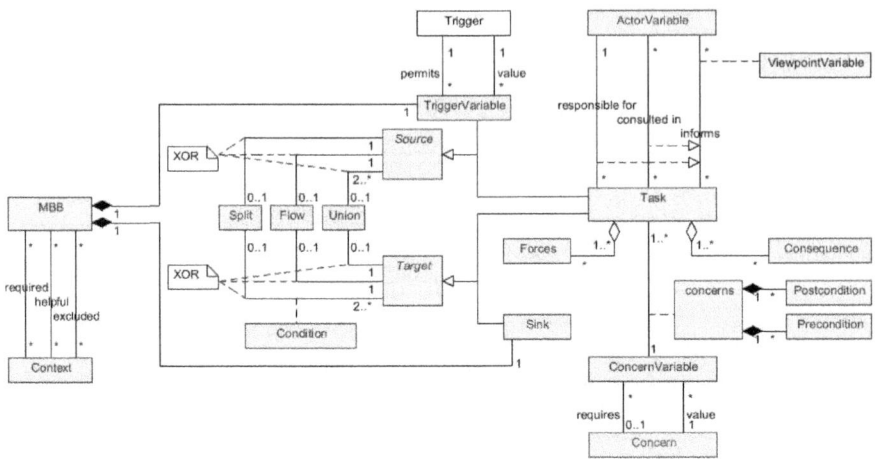

Fig. 6. Meta model of the method bulding blocks of the BEAMS approach

BEAMS actually distinguishes two subtypes of LBBs, namely information model building blocks IBBs and viewpoint building blocks VBBs. With the focus of this paper being on the MBBs, we abstain from giving in-depth information on these two types of building blocks and direct the interested reader both to Section 4, where examples of such building blocks are provided as well as to [7], where Buckl et al. discuss VBBs in more detail. Shortly summarizing the roles of these two types of building blocks, we may say that IBBs are used to define the *syntax* and *semantics* of the EA description language, i.e. to reflect the corresponding concern of the EA management function. VBBs are used to describe the language's *notation*[3], i.e. the way the EA-related information is presented.

Central to BEAMS is the notion of the MBB as re-usable solution for building an organization-specific EA management process mirroring the three phases of *develop and describe, communicate and enact* as well as *analyze and evaluate*. In this vein, we start with explaining the nature and inner organization of an MBB further taking into account the relationships to the other constituents of BEAMS. Figure 6 displays the constituents of an MBB and the relationship between these constituents. An MBB describes the different TASKs that are performed in order to achieve a certain goal under a given organizational CONTEXT. The MBB further specifies the ordering of the tasks and specifies SPLITs and UNIONs designating where tasks are alternative in their execution. For every SPLIT the MBB also describes the CONDITIONs that act during task execution. Complementing each MBB is started with a trigger represented in a TRIGGER VARIABLE. In configuring the EA management function this variable is filled with an actual TRIGGER, obeying the rules for doing so as supplied via the PERMITS relationship. Each task is executed by a corresponding actor represented by an

---

[3] The term "notation" is used in accordance with Kühn [30], whereas other publications refer to the notation as "concrete syntax".

ACTOR VARIABLE in the description of the method. The interplay of TASKs and FLOWs is described in a BPMN-like notation (cf. [10]) as exemplified in Figure 9. The notion of the "actor variable" is employed here to denote that the description of the MBB does not specify distinct actor or role in the using organization, but merely describes a responsibility of a person or group. Further, the MBB can specify that the actor variable is bound in respect to its organizational role, e.g. might express that an escalation based enactment mechanism only works, if a superordinate actor can be called upon. Beside to the mandatory relationship to the executing, i.e. *responsible* actor variable, each task may relate to other actor variables as well, namely variables representing actors that are *consulted* or *informed* during task execution. The distinction between the different levels of involvement pertaining to a single task is based on the RACI model of CobiT (see e.g. [27]), while a slightly different perspective is taken on the involvement level *informed*. For the purpose of describing MBBs, we assume that any actor involved in a task is informed, such that the responsible actor as well as consulted actors are counted as informed, too.

The *informed* relationship between a TASK and a corresponding actor (as represented in an ACTOR VARIABLE) is reified via a VIEWPOINT VARIABLE designating that the actor takes a specific viewpoint on the information relevant during the given task. The notion of the variable here again describes that the MBB does not make concrete prescriptions on the viewpoint to be used, but in turn allows to select an organization-specific viewpoint for accomplishing the task. Two remarks have to be added with respect to the VIEWPOINT VARIABLE. In the context of the BEAMS approach the concept of the viewpoint is discussed from a strongly notational perspective. This means that a viewpoint completely commits to the *syntax* and *semantics* specified by the underlying concern (IBB), while only the *notation* is specified in the viewpoint definition. This understanding of viewpoint is grounded in the work of Buckl et al. [8], who have discussed the relationships between viewpoints and concerns on a more formal basis. The notation of a viewpoint is specified via VBBs as transformation over the corresponding syntax via a pipe-based transformation language. Figure 7 shows an exemplary VBB defining a clustered visualization, where certain information objects describing the EA are converted to symbols (parameterization OUTER). Starting from these information objects the clustered visualization traverses a relationship (parameterization OUTER2INNER) to related information objects that are further converted to symbol (parameterization INNER). When exemplifying the approach in Section 4, we shall see more VBB-based transformations. A second remark pertains to the statement that an MBB does not make concrete prescriptions on the viewpoints to be used. While this is actually the case, an MBB may indeed recommend or discourage certain viewpoints, respectively. For example, a viewpoint variable used during an interview-related task may recommend textual viewpoints while on the contrary discouraging the utilization of viewpoints in the 'lines and boxes'-style. These statements nevertheless are no prescriptions on the level of the actual viewpoint to be used but rather on a meta-level, recommending or discouraging certain 'types' of viewpoints.

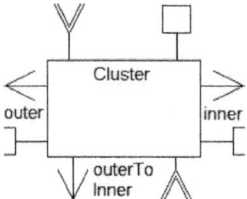

Fig. 7. VBB describing a clustered visualization

From an exterior perspective an MBB is associated with the architectural concern that its tasks cover. The concern specifies the area-of-interest in the enterprise on which the different steps taken in the MBB act. In line with the argumentation of Buckl et al. in [8] a concern may be identified with an information model, such that an MBB may during instantiation into a concrete process in the EA management function be parameterized with an according model. Nevertheless, many MBBs as e.g. ones associated with the activity of *develop and describe* can specify tasks without further knowledge on the actually associated area-of-interest. In this sense, an MBB does not directly link to a specific concern but to CONCERN VARIABLEs that are assigned with concrete concerns during configuration. Speaking more precisely, any MBB links to a CONCERN VARIABLE, specifying the MBB's or tasks *precondition* and *postcondition* concerning information demands. An exemplary precondition supplied as part of a concern variable would state that the MBB can only be executed, if information conforming to the given concern was documented. The same concern might conversely state as postcondition that the documented information was also cleared for communication. Any MBB may hence specify its dedicated set of pre- and postconditions, although these conditions must be specified using termini from the BEAMS terminology (or an extended version thereof). Finally, it has to be remarked that a CONCERN VARIABLE may specify a *lower bound*. By doing so the variable states a concern that is bound as value to the variable has to at least incorporate the concepts specified by the lower bound concern. Exemplifying this one should think of a task concerned with the assessment of projects. While no assumptions have to be made in respect to the exact area-of-interest during the assessment, it is nevertheless necessary for reasons of consistency to demand that the concern at least covers the project concept.

3.2 Development of BEAMS

The development of BEAMS demands due the DTN nature of the artifact input from related approaches. The very first approach concerned with EA management is the Zachman framework [53] dating back to 1987. Since that time, the number of researchers and practitioners targeting this area of interest has increased [32]. An overview on the current state-of-the-art in EA management is given by Aier et al. in [2] and the most active research groups in the area are determined by Schelp and Winter in [44]. We utilize the thereby identified 'major players' and their approaches to designing an EA management function in step one of the construction of BEAMS as input for the competing theories. Accordingly, the approaches of the following groups form the basis of our subsequent elaborations:

- EPFL Lausanne, Switzerland
- Novay, The Netherlands
- University of St. Gallen, Switzerland
- TU Berlin, Germany
- KTH Stockholm, Sweden
- TU Munich, Germany
- TU of Lisbon, Portugal

Step two involves analyzing the competing approaches identified in the first step following the method of hermeneutic text comprehension (cf. [54]) in order to determine their distinguishing characteristics. Thereby, we in particular focus on the essential goals of each approach and the respective means, i.e. processes, to achieve these goals. In this way, we identified the following goals:

1. reduce operating cost
2. increase disaster tolerance
3. reduce security breaches
4. ensure compliance
5. increase homogeneity
6. improve project execution
7. enhance strategic agility
8. improve capability provision
9. foster innovation
10. increase management satisfaction

Complementing, we identified different means to establish an organization-specific EA management function, e.g. an engineering based approach as presented by Aier et al. in [1], a pattern-based approach presented by Buckl et al. in [6,13], or an analysis-focused approach introduced by Johnson and Eksted in [28]. These different approaches or the contained methods represent the input for BEAMS. In the following the development of BEAMS, i.e. the instantiation of a DTN, is exemplified alongside the EAMPC of TU Munich that provides a catalog of best practices gathered from industry and academia and therefore can itself be regarded as a collection of competing design theories.

Following the idea of patterns as e.g. introduced by Alexander et al. in [4] different types of relationships between pattern may exists (cf. Noble in [39]). While patterns can provide alternative solutions, meaning they cannot be used in combination, i.e. represent competing solutions, other relationship types like *compatible, sub-, super-*, or *intersected* refer to patterns, which can be used in combination. Considering the patterns as contained in the EAMPC the different types of relationships as introduced above exists, especially within one type of patterns. These relationships should be considered in the construction of the

DTN for situational EA management, but are according to the design principle outlined above to be converted to formal relationships as far as possible. In the third step, we derive a number of assertions that are based on prominent characteristics of each approach as expressed in literature. For the patterns presented by Ernst in [16][4], for example, we formulated inter alia the following assumptions:

- Detailed information on applications and standardized technology needs to be available.
- A centralized IT organization is required to enable an architecture review process.
- Upper management support needs to be available to ensure architecture conformance of projects.

The assumptions formulated for the competing approaches are gathered and reformulated in order to use a common terminology. The following non-exhaustive list provides an overview on the thereby identified assumptions, which represent the organizational context descriptions of BEAMS:

- Centralized vs. decentralized IT organization
- Upper Management support for the EA management team
- EA management team has own budget, e.g. for architectural relevant project
- A dedicated tool for EA management is available or not
- Integration with other management function and processes, e.g. project portfolio management, is defined

The above identified goals of EA management and the organizational contexts are formulated in forms of conditions and mapped to the assumptions of the identified solutions. The suitability of the competing solutions for any combination of the conditions can then be defined utilizing a *fitting matrix* with the competing solutions on the y-axis, the identified conditions on the x-axis, and a scoring of the fitting function in the cell. The fitting function can thereby take a value form the set *required, excludes, helpful*. The patterns for enhancing standard conformity as proposed by Ernst in [16], for instance, would require a centralized IT organization, while the upper management support would only be helpful but is not necessarily required.

Based on this fitting matrix, a decision-making process for selecting one or more appropriate solutions for designing an situational EA management function is developed in step four. The appropriateness of the EA management function is heavily influenced by the goals pursued by the organization as well as by whatever pertinent issues are presented in the organizational context. Therefore, these constraints, i.e. goals and organizational context, determine whether a competing EA management approach succeeds or fails.

Finally, a technique supporting the utilization of BEAMS is developed in final step five. Thereby, the competing approaches, goals, organizational contexts,

---

[4] The patterns presented by Ernst in [16] represent an excerpt, which is also contained in the EAMPC [13].

a well as the process, which applies the fitting matrix, are reflected in the design of the technique. Possible realizations of the techniques may range from simple excel-based techniques in line with the scoring matrix of Pries-Heje and Baskerville (cf. [41]) to more sophisticated tools, which cannot only be used for selecting an appropriate EA management approach. Based on BEAMS, an organization-specific EA management functions can be constructed following the construction process of situational method engineering. Therefore, the following five steps have to be performed by the using organization.

- **Characterize situation:** The organizational context descriptions as introduced above have to be assessed and the goals of the EA management initiative have to be defined.
- **Tool-based assessment of method fragments:** A preselection and evaluation of the competing design theories contained in the DTN is returned by the tool, based on the provided information.
- **Selection of design theories:** The enterprise architect has to choose between the remaining theories or decide to use a combination.
- **Assembly of design theories:** The selected design theories need to be configured and adapted, e.g. regarding the ordering or the used terminology.
- **Establish situational EA management function:** The designed function has to be established, e.g. regarding governance structures, quality gates, etc.

Following the idea of *method administration* as discussed by Harmsen in [22], a performance measurement process should be set up that ensures sustainability of the EA management function. According to the typical management cycle as e.g. discussed by Deming in [14] or Shewart in [46] a governance function should be established that measures the achievement of objectives and if necessary adapts the EA management function accordingly by reentering the above presented configuration process. Furthermore, an extension mechanism needs to be implemented in the DTN for situational EA management in order to integrate new or update existing design theories if necessary.

4   Exemplifying BEAMS – Designing an Organization-Specific EA Management Function

In this section we describe an exemplary application of the BEAMS in a fictional organization, namely the financial service provider BSM.

The situation in respect to the organizational context of the EA management function of BSM can be characterized as follows: Over the years BSM purchased different other financial service providers, adapted their business processes, and incorporated their business applications. This has lead to a complex and highly heterogeneous application portfolio that BSM has difficulties in evolving and maintaining. In order to increase maintainability of its business applications, BSM wants to reduce their total number. For doing so, the organization decides to launch an EA management pilot project. With the organizational structure of BSM being grown over a series of acquisitions of other financial service providers, the organization has retained a number of independent IT units and hence a 'decentralized IT'. The EA management pilot is driven by a small EA management team located in a staff unit of the CIO's office. While this means that the EA management initiative can rely on high-level management support, especially the decentralized structure of the IT departments makes it necessary to use the pilot project for marketing and illustrating the benefit of the new and overarching management function. In addition, the EA management team has to deal with missing tool support for EA management as currently no further budget for the pilot is available.

Not aiming too high, BSM sets the goal of the EA management pilot to 'increased transparency' focused on the concern 'business applications used by organizational units'. This forms the initial EA-related problem that BSM seeks to address by the EA management initiative. This problem statement is covered by an IBB presented by BEAMS. This IBB comprises an information model that contains two classes BUSINESS APPLICATION and ORGANIZATIONAL UNIT as shown in Figure 8.

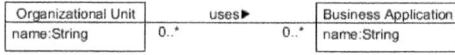

Fig. 8. Initial information model of BSM

Based on the identified organizational contexts an assessment of the BBs of BEAMS is performed, e.g. MBBs which build on the availability of an EA management tool are excluded from the selection process. The EA management team of BSM understands that the application owners would be pleased to deliver the information about their application's use in different organizational units to demonstrate the importance of the application they are responsible for. In contrast, stakeholders from the business departments may keep distance but can get in first contact with the EA management initiative. They decide to interview the application owners and according business departments to gather the corresponding information. The corresponding MBB is selected as it states (as CONSEQUENCE) to be beneficial for marketing, which can be performed prior to the interview by illustrating the objectives of the endeavor. The MBB shown in Figure 9 describes the steps of conducting interviews to gather information conforming to a concern. As a post-condition the MBB states that the corresponding concern is *documented* meaning the according information is available. Further, the MBB can be selected for the purpose that BSM wants to pursue, as interviews are well-suited to document current state EAs. Would the problem statement conversely aim at target or planned states of the EA, different MBBs had to be taken.

The MBB selected before states as a consequence that the application of the method may result in inconsistent information. Put in the context of our example, an application owner could state a business department, which according to his knowledge uses the respective business application but the corresponding

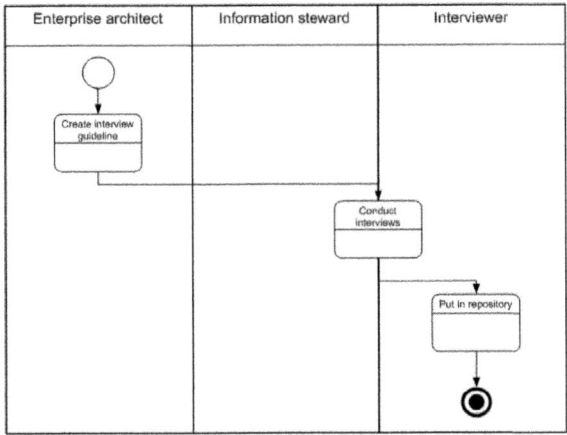

Fig. 9. MBB for gathering EA information via interviews

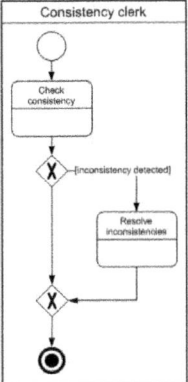

Fig. 10. MBB for checking consistency in the gathered information

business department denies that. In order to overcome this drawback of inconsistent information, another develop & describe-related MBB admissible in the given circumstances is available, which is concerned with tasks for consistency checking in EA documentations. The EA management team of BSM decides to make use of this MBB in order to ensure consistency in the documentation. The consistency-checking MBB as shown in Figure 10 takes a documented concern in its pre-condition and states that after the execution of the corresponding tasks, the concern, more precisely the corresponding information, is *consistent*.

BSM selects a communicate & enact MBB, which publishes the gathered information describing the current status of the EA via publishing it in the corporate intranet. Figure 11 illustrates the assembled method of BSM. During assembling the MBBs, the stakeholder variables are replaced by organization specific actors and roles, i.e. the 'information steward' is replaced by the 'application owners'

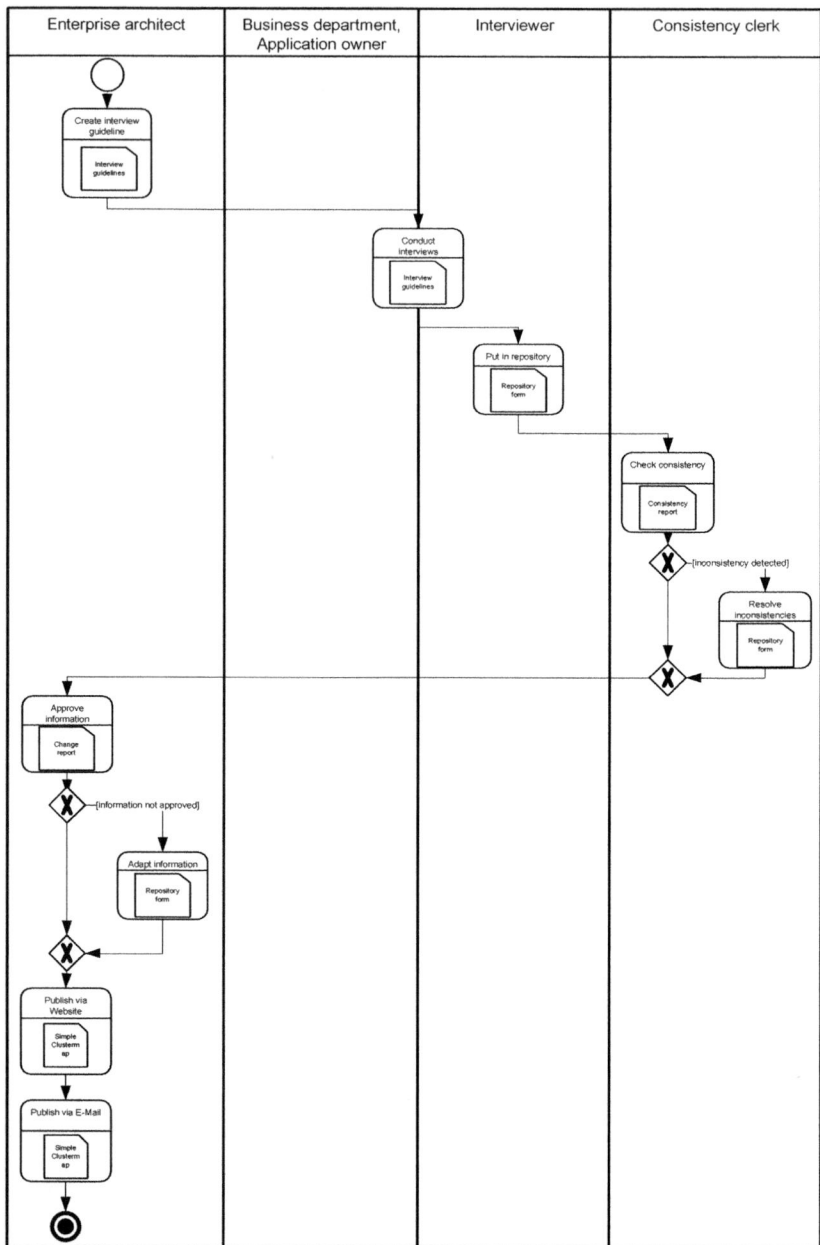

Fig. 11. Organization-specific method for increasing transparency

and 'business departments' respectively. Furthermore, the viewpoints utilized during the execution of the method are further specified utilizing VBBs as reports, forms, and maps. For the 'publish via website' task for example, the VBB of a 'simple cluster map' is used. A graphical description of the VBB for a simple cluster map according to the information model given in Figure 8 is illustrated in Figure 12, while the resulting view is shown in Figure 13.

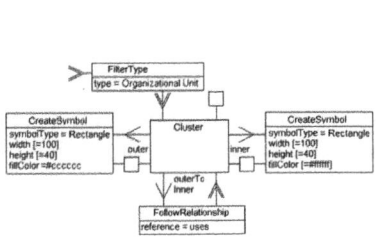

Fig. 12. VBB simple clustermap

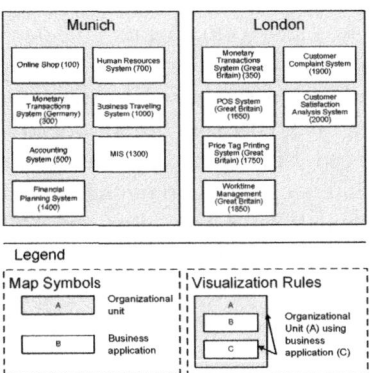

Fig. 13. Simple clustermap ov BSM

Finally, BSM has to establish their organization-specific management function based on BBs of BEAMS. Therefore, the roles newly introduced by the EA management endeavor, e.g. the consistency clerk, have to be filled and the initial information gathering procedure has to be performed. As suggested in Section 3 a performance measurement process is set up. After publishing the current state of the EA via the corporate intranet, the CIO as the stakeholder of the problem statement as well as the business departments and the application owners as the participating users are asked for their satisfaction with the achieved results. Due to the positive feedback, the CIO decides to extend the scope of the EA management function. The goal of the EA management initiative is changed to 'increase homogeneity' of the application landscape. The concern is consequently extended, i.e. a superconcern in the sense of Buckl et al. [9] is used. This complementary leads to an extended information model for BSM to include the 'technologies used by a business application' as well as possibilities to define a 'technology as standard'. The corresponding IBB is illustrated in Figure 14.

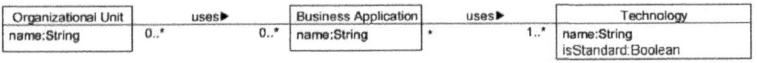

Fig. 14. Information model for increasing homogeneity

Performing further adaptations of the EA management function of BSM, the design team revisits the method framework of BEAMS (cf. Figure 4). In order to pursue the goal 'increase homogeneity' additional information about the organization's EA has to be gathered. In order to keep the investment for gathering the additional information low, the EA management team decides to send around questionnaires, in which the application owners are requested to state the technologies that their corresponding application uses. A corresponding MBB is used to incorporate this information. Based on this information an expert team assesses the technologies and decides on their standard conformity. More precisely, the the expert team decides which technologies should be supported as standards. In this sense, the EA management function is extended to incorporate activities for *analyzing and evaluating*, more precisely an MBB for this purpose. Finally, the EA management team decides to adapt the communication mechanisms by introducing an additional viewpoint conveying standardization relevant information. For doing so a specialized version of the cluster map is created decorating the symbol creation rules for the application with a color-coding indicating, whether all related technologies are standard-conform or not. Bringing together the different BBs a revised version of the EA management function as shown in Figure 15 is developed.

5   Related Work

The approach of BEAMS is not at lest due to his DTN nature different from the other approaches fro EA management as found in literature. But the DTN nature also explains the intricate relationship between BEAMS and different EA management approaches, from which BEAMS draws its proven-practice building blocks. Against that background a detailed comparison of BEAMS with the corresponding approaches is likely to fall short of any novel insights. We nevertheless take two selected 'traditional' EA management approaches as reference points of comparison (see Section 5.1) to highlight the different nature of BEAMS. Complementing these comparisons we further summarize two *contingency*-based approaches for EA management and show in Section 5.2 how these relate to BEAMS.

5.1   Traditional EA Management Approaches

The Open Group Architecture Framework (TOGAF) (cf. [48]) is perhaps the most well-known framework for EA management. In its most recent version 9.0, TOGAF presents both an architecture development method (ADM) and an information model for architectural description, the so called "enterprise content metamodel". The cyclic ADM is designed as reference method for performing an "architecture project", which in the sense of TOGAF is the natural way of performing EA management. Such architecture project is set up in a preliminary phase that decides over scope and reach of the project with respect to the EA, but also over the utilization of tools. Further, decisions are taken over the linkage of the EA management project to existing enterprise-level management processes, as e.g. business object management. During the first phsae of project execution, the reach and scope are further concretized via a selection of

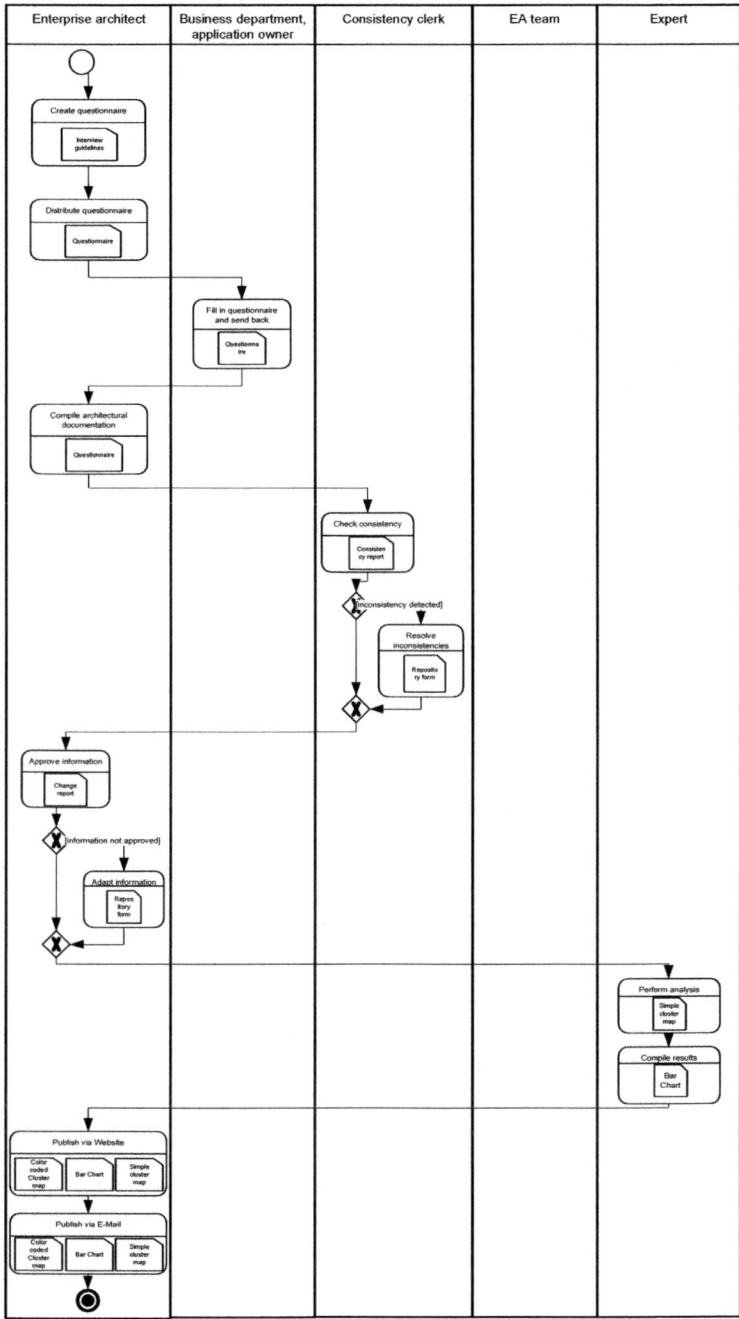

Fig. 15. Organization-specific method for increasing homogeneity

the corresponding stakeholders. Additionally, goals and requirements of the EA management project are defined making concrete prescriptions on the EA vision to pursue. In this sense the first two phases of the ADM relate to the activity of *configure and adapt* as described in the method framework of BEAMS (cf. Section 3). The EA vision is further detailed in three subsequent phases of the TOGAF ADM, namely the *business architecture, information systems architecture* and *technology architecture* development. During these phases the current state of the architecture is documented, a target state is developed, and gap analyses between these states are performed. In line with the prescriptions of TOGAF the documentation of the current state has to developed prior to the development of the target one. In the next two phases of the ADM (*opportunities and solutions* and *migration planning*) the results of the gap analysis are used to propose and plan transformation activities that change the overall EA. The implementation of this change is monitored in the *implementation and governance* phase, whereas the final phase *architecture change management* is concerned with assessing the overall performance of the EA management project. Complementing the description of the activities, TOGAF describes the input and output artifacts of the different phases, although the thereby utilized visualizations are only informally tied to the underlying information model. This content metamodel is presented as an extensible model centered around a monolithic core. The provided extensions introduce additional concepts such as KPIs or goals into the model, but no prescriptions are made, when to use which extension. Further, the information model remains on a rather abstract level abstaining from details as datatypes, multiplicities or constraints that would nevertheless be needed to ensure model consistency. Due to the project nature of TOGAF's EA management only a few remarks a spent on the establishment of a continuous EA management function, which is further reflected by the fact that maintenance activities for the EA documentation and continuous communication mechanisms are not discussed.

A very frequently quoted academic approach to EA management is the archimate approach presented by Lankhorst et al. in [33]. Central to this approach are the activities of documentation, communication and analysis of EAs. These activities are accounted for by introducing a specialized modeling language, the archimate modeling language, based on an information model (cf. e.g. [29]) covering the three facets *structure, behavior,* and *information* on the different architectural layers *business, application,* and *infrastructure.* This information model, while being monolithic in design, partially accounts for the diversity in the organizations understanding of their EAs by support 'short-cut' modeling. This means that, although the model assumes three distinct architectural layers, the layering is not strict, but intermediary concepts may be omitted, if necessary. This built-in flexibility of modeling does nevertheless not come without cost, but can lead to imprecise and inconsistent models, especially if concepts on an intermediary layer are added later. The lack of prescriptions in the field of model adaptation does not prevent the approach from giving comprehensive insights into both visualization and analysis techniques building on these models. The archimate approach presents different graphical notations for visualizing architectural information, as e.g. a "business support map". Further, different quantitative analysis techniques building on the archimate meta-model are described (see e.g. [25]). Nevertheless, when it comes to prescriptions on how to adapt the archimate approach to a specific organization, literature becomes scarce and actual prescriptions are missing.

5.2   Contingency Approaches to EA Management

Only recently *contingency approaches* to EA management have gained some prominence as means to account for the organization-specificity of an EA management function. An early example of such approach is presented by Leppänen et al. in [34]. Central to their approach is the "EACon" framework, an organized collection of contingency factors of EA method engineering. This framework builds on rich literature in the field of EA management and devises the central contingency factors as found there, namely "EA method goals", "enterprise" as well as "environment", and "roles" as well as "resources". These factors may well be identified either with the EA-related goals and the contextual factors of BEAMS, or with system of actor variables introduced therein. Leppänen et al. further detail on a contingency factor called "EA management" that is concerned with decisions rights and coordination means of EA management which are conversely covered by the actor modeling of BEAMS. Other aspects as "communication means" are only briefly alluded to in the contingency framework reflecting the strongly method centric perspective taken therein. This further reverberates in the rather short discussion on the concern of EA management showing that language aspects are not discussed by Leppänen et al. in [34]. Relating the work to the BEAMS approach, the contingency framework may well be used to structure and to organize the contextual influences that pertain to the approaches used to build BEAMS. When it nevertheless comes to concrete prescriptions or contingencies on aspects of EA description, the work of Lepänen et al. [34] may only serve as an abstract reference point providing possible dimensions of adaptation.

In [42], Riege and Aier outline a contingency approach to EA management with emphasis on the method aspect of the management activity. This is primarily reflected in discussion on the organizational setting that may constrain the implementation of an EA management function or project. Complementing these discussions, the approach presents abstract goal-like statements that may be helpful to frame the goal of the overall EA management activity. Concrete prescriptions on the method steps to be taken or actual EA-related goals for practical settings are conversely out of the scope of the contingency approach. The same is true, when it comes to language aspects, although Riege and Aier add a short side node on the "constitution of the EA", which should adequately reflected as documentation fed to the management activities. In this sense, the BEAMS approach may be seen as stringent continuation of the work of Riege and Aier as presented in [42] accounting for both sides of the EA management coin, management methods and description languages.

6   Outlook

This article contributes a building block based approach to the field of EA management governance. With the BEAMS an organization seeking to establish a specific EA management function should be able to leverage operational and practice-proven design prescriptions. In this sense, the presented approach continues the work started by the contingency based EA management approaches of Leppänen et al. in [34] as well as Riege and Aier in [42]. With its grounding in the EAM pattern approach of Buckl et al. (cf [6,13]) and other practice-proven approaches from academic research as well as from standardization bodies, BEAMS provides a comprehensive approach to a highly relevant topic of information systems research and practice. With all the contributing approaches being successful in practice, one can sensibly assume that BEAMS is applicable in different practical settings, although a thorough evaluation on this topic is yet to be undertaken. First practice projects currently implemented building on the prescriptions of BEAMS nevertheless are developing in promising ways.

A future challenge in the context of BEAMS is associated with the evolvement of the approach itself. Up to this point, BEAMS is initialized with input from various sources, but as EA management-related research continues, future findings and results may provide a valuable addition to the approach. The method for constructing a DTN instantiation as presented by Pries-Heje and Baskerville (cf. discussions in Section 2.1) may be helpful in this context, but is by nature limited to one stream of evolution. In contrary we expect that BEAMS will attract a similar community as the EA management pattern catalog (cf. [13]) did, such that methods, mechanisms, and techniques for collaboratively evolving the knowledge base of BEAMS are needed. This especially applies with respect to the intermediary artifacts of the patterns and case descriptions that were used for constructing the initial set of building-blocks. The notion of these intermediary artifacts further relates to another future challenge – the tool support.

The development of an organization-specific EA management function is – even with the prescriptions and building-blocks provided by BEAMS – a complex task. The consistent integration of the MBBs keeping track of the pre- and post-conditions, respectively, requires careful attention. The same also applies for the integration of the information models and viewpoints that build the LBBs used in the EA management function. With this background and intricate interrelations, the design activity for organization-specific EA management functions calls for tool support. A building-block based a configurator for EA management functions may further be helpful as vehicle for evolving the knowledge base of BEAMS, as the configurations and adaptations made in such tool may be technically reflected to the knowledge base and analyzed using statistical means. On the contrary, the configuration tool must provide mechanisms to export the configured EA management function as a description file.

The work [36] of Matthes et al. describes that current tools for supporting EA management can be categorized into two large groups of tools, namely *meta-modeling* tools of high flexibility and *methodology-driven* tools delivering a pre-defined information model and method framework. Tools of the latter type may be regarded as 'traditional' EA management approaches in the sense of the discussion from Section 5.1, whereas tools of the former type may be used to implement arbitrary EA management approaches. In this sense, meta-modeling tools may interpret the description file exported from the BEAMS configurator. This conversely calls for a standardization of the exchange and configuration file format to facilitate the re-use of the defined configuration.

References

1. Aier, S., Kurpjuweit, S., Saat, J., Winter, R.: Enterprise Architecture Design as an Engineering Discipline. AIS Transactions on Enterprise Systems 1, 36–43 (2009)
2. Aier, S., Riege, C., Winter, R.: Unternehmensarchitektur – Literaturüberblick Stand der Praxis. Wirtschaftsinformatik 50(4), 292–304 (2008)
3. Alexander, C.: The Timeless Way of Building. Oxford University Press, New York (1979)
4. Alexander, C., Ishikawa, S., Silverstein, M., Jacobson, M., Fiksdahl-King, I., Angel, S.: A Pattern Language. Oxford University Press, New York (1977)
5. Buckl, S., Ernst, A.M., Lankes, J., Matthes, F.: Enterprise Architecture Management Pattern Catalog (Version 1.0, February 2008). Technical report, Chair for Informatics 19 (sebis), Technische Universität München, Munich, Germany (2008)
6. Buckl, S., Ernst, A.M., Lankes, J., Schneider, K., Schweda, C.M.: A pattern based approach for constructing enterprise architecture management information models. In: Wirtschaftsinformatik 2007, Karlsruhe, Germany, pp. 145–162. Universitätsverlag Karlsruhe (2007)
7. Buckl, S., Gulden, J., Schweda, C.M.: Supporting ad hoc analyses on enterprise models. In: 4th International Workshop on Enterprise Modelling and Information Systems Architectures (2010)
8. Buckl, S., Krell, S., Schweda, C.M.: A formal approach to architectural descriptions – refining the iso standard 42010. In: 6th International Workshop on Cooperation & Interoperability – Architecture & Ontology, CIAO 2010 (2010)
9. Buckl, S., Matthes, F., Schweda, C.M.: Interrelating concerns in ea documentation – towards a conceptual framework of relationships. In: 2nd European Workshop on Patterns for Enterprise Architecture Management (PEAM 2010), Paderborn, Germany (2010)
10. Buckl, S., Matthes, F., Schweda, C.M.: A modeling language for describing ea management methods. In: Modellierung betrieblicher Informationssysteme (MobIS 2010) (2010)
11. Buckl, S., Matthes, F., Schweda, C.M.: Towards a method framework for enterprise architecture management – a literature analysis from a viable system perspective. In: 5th International Workshop on Business/IT Alignment and Interoperability (BUSITAL 2010) (2010)
12. Buschmann, F., Meunier, R., Rohnert, H., Sommerlad, P., Stal, M.: Pattern-oriented software architecture: a system of patterns. John Wiley & Sons, Inc., New York (1996)
13. Chair for Informatics 19 (sebis), Technische Universität München. Eam pattern catalog wiki (2010), http://eampc-wiki.systemcartography.info (cited 2010-07-01)
14. Deming, E.W.: Out of the Crisis. Massachusetts Institute of Technology, Cambridge (1982)

15. Eisenhardt, K.M., Graebner, M.E.: Theory building from cases: Opportunities and challenges. Academy of Management Journal 50(1), 25–32 (2007)
16. Ernst, A.: Enterprise architecture management patterns. In: PLoP 2008: Proceedings of the Pattern Languages of Programs Conference 2008, Nashville, USA (2008)
17. Ernst, A.M.: A Pattern-Based Approach to Enterprise Architecture Management. PhD thesis, Technische Universität München, München, Germany (2010)
18. Frank, U.: Multi-perspective enterprise modeling (memo) – conceptual framework and modeling languages. In: Proceedings of the 35th Annual Hawaii International Conference on System Sciences (HICSS 2002), Washington, DC, USA, pp. 1258–1267 (2002)
19. Gamma, E., Helm, R., Johnson, R., Vlissides, J.M.: Design Patterns: Elements of Reusable Object-Oriented Software. Addison-Wesley Professional Computing Series. Addison-Wesley Professional, Munich (1994)
20. Gehlert, A., Schermann, M., Pohl, K., Krcmar, H.: Towards a research method for theory-driven design research. In: Hansen, H.R., Karagiannis, D., Fill, H.G. (eds.) Business Services: Konzepte, Technologien, Anwendungen, 9, Wien, Austria. Internationale Tagung Wirtschaftsinformatik, vol. 1, pp. 441–450. Österreichische Computer Gesellschaft (2009)
21. Hafner, M., Winter, R.: Vorgehensmodell für das management der unternehmensweiten applikationsarchitektur. In: Ferstl, O.K., Sinz, E.J., Eckert, S., Isselhorst, T. (eds.) Wirtschaftsinformatik, pp. 627–646. Physica-Verlag, Heidelberg (2005)
22. Harmsen, A.F.: Situational Method Engineering. PhD thesis, University of Twente, Twente, The Netherlands (1997)
23. Henderson, J.C., Venkatraman, N.: Strategic alignment: leveraging information technology for transforming organizations. IBM Systems Journal 38(2-3), 472–484 (1999)
24. Hevner, A.R., March, S.T., Park, J., Ram, S.: Design science in information systems research. MIS Quarterly 28(1), 75–105 (2004)
25. Iacob, M.-E., Jonkers, H.: Quantitative analysis of enterprise architectures. In: Konstantas, D., Bourrières, J.-P., Léonard, M., Boudjlida, N. (eds.) Interoperability of Enterprise Software and Applications, Geneva, Switzerland, pp. 239–252. Springer, Heidelberg (2006)
26. International Organization for Standardization. ISO/IEC 42010:2007 Systems and software engineering – Recommended practice for architectural description of software-intensive systems (2007)
27. IT Governance Institute. Framework Control Objectives Management Guidelines Maturity Models (2009), http://www.isaca.org/Knowledge-Center/cobit (cited 2010-06-18)
28. Johnson, P., Ekstedt, M.: Enterprise Architecture – Models and Analyses for Information Systems Decision Making, Studentlitteratur, Pozkal, Poland (2007)
29. Jonkers, H., van Burren, R., Arbab, F., de Boer, F., Bonsangue, M., Bosma, H., ter Doest, H., Groenewegen, L., Scholten, J., Hoppenbrouwers, S., Iacob, M.-E., Janssen, W., Lankhorst, M., van Leeuwen, D., Proper, E., Stam, A., van der Torre, L., van Zanten, G.: Towards a language for coherent enterprise architecture descriptions. In: 7th International Enterprise Distributed Object Computing Conference (EDOC 2003), Brisbane, Australia. IEEE Computer Society, Los Alamitos (2003)
30. Kühn, H.: Methodenintegration im Business Engineering. PhD thesis, Universität Wien (2004)

31. Kurpjuweit, S., Winter, R.: Viewpoint-based meta model engineering. In: Reichert, M., Strecker, S., Turowski, K. (eds.) Enterprise Modelling and Information Systems Architectures – Concepts and Applications, Proceedings of the 2nd International Workshop on Enterprise Modelling and Information Systems Architectures (EMISA 2007), St. Goar, Germany, Bonn, Germany, October 8-9. LNI, pp. 143–161. Gesellschaft für Informatik (2007)

32. Langenberg, K., Wegmann, A.: Enterprise architecture: What aspect is current research targeting? Technical report, Laboratory of Systemic Modeling, Ecole Polytechnique Fédérale de Lausanne, Lausanne, Switzerland (2004)

33. Lankhorst, M.: Enterprise Architecture at Work: Modelling, Communication and Analysis. Springer, Heidelberg (2005)

34. Leppänen, M., Valtonen, K., Pulkkinen, M.: Towards a contingency framework for engineering and enterprise architecture planning method. In: 30th Information Systems Research Seminar in Scandinavia (IRIS), pp. 1–20 (2007)

35. Luftman, J.N., Lewis, P.R., Oldach, S.H.: Transforming the enterprise: The alignment of business and information technology strategies. IBM Systems Journal 32(1), 198–221 (1993)

36. Matthes, F., Buckl, S., Leitel, J., Schweda, C.M.: Enterprise Architecture Management Tool Survey 2008. Chair for Informatics 19 (sebis), Technische Universität München, Munich, Germany (2008)

37. McDermin, J.A.: Software engineer's reference book. Butterworth Heinemann, Oxford (1994)

38. Niemann, K.D.: From Enterprise Architecture to IT Governance – Elements of Effective IT Management. Vieweg+Teubner, Wiesbaden (2006)

39. Noble, J.: Classifying relationships between object-oriented design patterns. In: Australian Software Engineering Conference (ASWEC), pp. 98–107. IEEE Computer Society, Los Alamitos (1998)

40. Nunamaker, J.F., Chen, M., Purdin, T.D.M.: Systems development in information systems research. J. Manage. Inf. Syst. 7(3), 89–106 (1991)

41. Pries-Heje, J., Baskerville, R.: The design theory nexus. MIS Quarterly 32(4), 731–755 (2008)

42. Riege, C., Aier, S.: A contingency approach to enterprise architecture method engineering. In: Service-Oriented Computing - ICSOC 2008 Workshops. LNCS, vol. 5472, pp. 388–399 (2009)

43. Schekkerman, J.: Enterprise Architecture Good Practices Guide – How to Manage the Enterprise Architecture Practice. Trafford Publishing, Victoria (2008)

44. Schelp, J., Winter, R.: On the interplay of design research and behavioral research – a language community perspective. In: Proceedings of the Third International Conference on Design Science Research in Information Systems and Technology (DESRIST 2008), Westin, Buckhead, Atlanta, Georgia, USA, Georgia State University, Atlanta, May 7-9, pp. 79–92 (2008)

45. Schönherr, M.: Towards a common terminology in the discipline of enterprise architecture. In: Aier, S., Johnson, P., Schelp, J. (eds.) Pre-Proceedings of the 3rd Workshop on Trends in Enterprise Architecture Research, Sydney, Australia, pp. 107–123 (2008)

46. Shewhart, W.A.: Statistical Method from the Viewpoint of Quality Control. Dover Publication, New York (1986)

47. Standish Group International. The chaos report 2009 (2009), http://www.standishgroup.com/newsroom/chaos_2009.php (cited 2010-02-25)

48. The Open Group: TOGAF "Enterprise Edition" Version 9 (2009),
    http://www.togaf.org (cited 2010-02-25)
49. van Aken, J.E.: Management research based on the paradigm of the design sciences:
    The quest for field-tested and grounded technological rules. Journal of Management
    Studies 41(2), 219–246 (2004)
50. van den Berg, M., van Steenbergen, M.: Building an Enterprise Architecture Prac-
    tice – Tools, Tips, Best Practices, Ready-to-Use Insights. Springer, Dordrecht
    (2006)
51. Wegmann, A.: The Systemic Enterprise Architecture Methodology (SEAM). Tech-
    nical report, EPFL (2002)
52. Wegmann, A., Balabko, P., Lê, Regev, G., Rychkova, I.: A method and tool for
    business-it alignment in enterprise architecture. In: Proceedings of the CAiSE 2005
    Forum, Porto, Portugal, pp. 113–118 (2005)
53. Zachman, J.A.: A framework for information systems architecture. IBM Syst.
    J. 26(3), 276–292 (1987)
54. Zelewski, S.: Erkenntnisinstrumente der betriebswirtschaftslehre. In: Betrieb-
    swirtschaftslehre – Band1, München, Germany, Oldenbourg (2008)

Organizational and Design Engineering of the Operational and Support Components of an Organization: The Portuguese Air Force Case Study

Carlos Páscoa[1,2] and José Tribolet[2,3]

[1] Department of University Education, Portuguese Air Force Academy,
Granja do Marquês, Sintra, Portugal
[2] Department of Information Systems and Computer Science,
Instituto Superior Técnico, Technical University of Lisbon, Portugal
[3] CODE - Center for Organizational Design & Engineering, INOV,
Rua Alves Redol 9, Lisbon, Portugal
cjpascoa@gmail.com, jose.tribolet@inesc.pt

Abstract. The general organization of today results from a combination of elements that makes it a very complex entity, in which the operational and support dimensions should co-exist in a dynamic and constant balance, whose configuration must have flexible and adaptable mechanisms to the outside world. The operational dimension of the organization, in this context, performs a key role because it is linked to the executables that generate output to the exterior, representing the added value and allowing it to achieve measurable objectives. Business Processes perform a key role and are essential for ensuring the availability of resources for proper organization functioning. As processes increase in complexity, it is essential to identify, given the complexity of procedures, what is the relationship between the operational component (generator of value) and the support component and, also, how to draw, organize and manage an organization, in the human and material resources domain, considering i) multiple restrictions; ii) critical needs of real time; iii) various configurations. The Portuguese Air Force, based on a coherent set of principles, initiated a process of change. The core business, flying, is a proven success and the evolving principles can be used in the Organization itself to improve self-awareness.

Keywords: Organizational and Design Engineering, Business Architecture, Information Systems Architecture, Information Systems, Business Strategy and Information Systems alignment, Operational Activity, Air Force Mission, Maintenance Activity, Aircraft Operational Use, Strategic, Operational and Tactic Management, Systems Theory.

1 Introduction

The current organization, in its strategic, tactical and operational components results from a combination of elements that makes it a very complex entity, in which the operational and support dimensions should co-exist in a dynamic and constant balance.

F. Harmsen et al. (Eds.): PRET 2010, LNBIP 69, pp. 47–77, 2010.

To attain the desired balance, flexible and adaptable mechanisms must be created and maintained. The operational component of the organization, in this context, performs a key role because it is linked to the executables that generate output to the exterior, representing the added value and allowing it to achieve measurable objectives.

To attain its purposes, the organization manages entities acting on business processes that, in turn, are used by other business processes. These processes are essential for ensuring the availability of resources for proper organization functioning. As processes increase in complexity, being used by more actors and other processes, each with a set of constraints, it becomes more difficult to manage the organization, almost in real time, in its many dimensions and configurations. Therefore, it is necessary to create and adapt transformation mechanisms, that reducing complexity, are able to maintain the necessary balance for the operation to succeed.

In this context, it is therefore essential to identify, given the complexity of procedures, what is the relationship between the operational component (generator of value) and the support component and, also, how to draw, organize and manage an organization, in the human and material resources domain, considering i) multiple restrictions ii); critical needs of real time iii); various configurations.

This paper shows how the Portuguese Air Force initiated a change process, using coherent principles of Organizational and Design Engineering (ODE), taking as tools proven concepts associated to flying (like the organizational cockpit), towards situational awareness.

The main objective, however, is not to describe the change process but, instead, to present some of its results in terms of organization artifacts that are helping the Air Force to improve its situational awareness.

Due to the size of the paper, it is not possible to describe, in detail, all the artifacts and their real impact in the Air Force life. Therefore, the aim is to explain briefly:

- What the overall objective is;
- The purpose of each artifact, and
- How the flying concepts helped to discipline actions taken.

Section 2 presents the concepts associated with change: General Systems Theory, ODE, Business Motivation Model, Self-awareness, Flexibility, Agility and Change. Section 3 describes some principles and concepts associated with flying such as personnel and mission preparation. Section 4 briefly introduces the change process (that is not the core objective of this paper), and Section 5 presents the core objective of this paper as it describes concept application to the Organization. Section 6 presents the conclusion.

2 Literature Review

In order to understand some of the organizational paradigms, this paragraph is divided into several sections. Each of the sections addresses an important related organizational issue.

New forms of designing and engineering the organization are presented by communities that struggle to transform enterprise engineering into a steady and robust science.

Thus, in order to try to get an idea of the work done in areas of interest, the related research is structured in the following way:

- General Systems Theory. Describes important elements related to system's simple forms and basic relations.
- Organizational and Design Engineering. Presents theoretical foundations for the organization self-awareness taking in consideration technical and social aspects.
- Enterprise Architecture. Processes, Skills and Views of the Organization, defines a set of items necessary to understand the organization's business, particularly those that contribute to the identification of the Information Systems Architecture (ISA), the architectural sub-components, models and views of the organization.
- Enterprise Governance. A recent term that introduces the governance perspective, which includes the strategic and operational performance with focus on compliance and performance concerns.
- Business Motivation Model. Provides a scheme or structure for developing, communicating, and managing business plans in an organized manner. This paragraph describes the model's concepts and relations.
- Self-awareness, Flexibility and Change. Key factors in any organization, which drive the ability to identify and store the organization's knowledge are presented as decisive for attaining situation awareness and near time reaction, thus providing competitive advantage.
- Changing the Organization. To become self-aware, once understood all the concepts behind adaptability and flexibility the organization must delineate a path to find the TO BE. Change theories provide known forms of achieving change.

2.1 General Systems Theory

The General Systems Theory (GST) is based on three basic premises: i) systems exist within systems; ii) systems are opened; iii) systems functions depend on its structure.

Fredrick Hegel (1770-1831) formulated four statements concerning the nature of systems, stating that:

- The whole is more than the sum of the parts.
- The whole defines the nature of the parts.
- The parts cannot be understood by studying the whole.
- The parts are dynamically interrelated or interdependent.

In what concerns its constitution, systems can be concrete, conceptual, abstract or unperceivable. The concrete system (living or non-living) is real in the dimensions of space and time and is defined as consisting of at least two units or objects. A standard biological definition uses the following characteristics: Self-regulation; Organization; Metabolism and Growth; Reaction Capacity; Adaptability; Reproduction Capability and Development Capability.

Complex, organized and open systems are also characterized by its capacity for autopoiesis, a theory created by H. Maturana and V. Varela (1974) [1], which means "self-renewing" allowing living systems to be autonomous.

Today, there is a near total agreement on which associated properties together comprise a general systems theory of open systems. Ludvig von Bertalanffy (1955) [2], Joseph Litterer (1969) and other distinguished people have formulated the hallmarks of such a theory. The list below sums up their efforts in identifying principles for characterizing relations between systems.

- *Interrelationship and interdependence* of objects and their attributes: Unrelated and independent elements can never constitute a system.
- *Holism*: Holistic properties not possible to detect by analysis should be possible to define in the system.
- *Goal seeking*: Systemic interaction must result in some goal or final state to be reached, or some equilibrium point being approached.
- *Transformation process*: All systems, if they are to attain their goal, must transform inputs into outputs. In living systems, this transformation is mainly of a cyclical nature.
- *Inputs and outputs*: In a closed system, the inputs are determined once and for all; in an open system additional inputs are admitted from its environment.
- *Entropy*: This is the amount of disorder or randomness present in any system. All non-living systems tend toward disorder; left alone they will eventually lose all motion and degenerate into an inert mass. When this permanent stage is reached and no events occur, maximum entropy is attained.
- *Regulation*: The interrelated objects constituting the system must be regulated in some fashion, so that its goals can be accomplished. Regulation implies that necessary deviations will be detected and corrected. Feedback is therefore a requisite of effective control. Typical of surviving open systems is a stable state of dynamic equilibrium.
- *Hierarchy*: Systems are generally complex wholes made up of smaller subsystems. This nesting of systems within other systems is what is implied by hierarchy.
- *Differentiation*: In complex systems, specialized units perform specialized functions. This is a characteristic of all complex systems and may be called specialization or division of labor.
- *Equifinality and multifinality*: Open systems have equally valid alternative ways of attaining the same objectives from different initial conditions (convergence) or, from a given initial state, obtain different, and mutually exclusive, objectives (divergence).

2.2 Organizational and Design Engineering

Universities and Institutes in the civilian world are currently going deep into investigating the integration of knowledge coming from different fields and paradigms. One of these fields of interest is Organizational Design and Engineering (ODE), a multidisciplinary research project born at the Department of Computer Science and Engineering of the Instituto Superior Técnico in Lisboa, Portugal, and now established at the Centre for Organizational Design and Engineering (CODE).

ODE is defined as "the application of social science and computer science research and practice to the study and implementation of new organizational designs, including the integrated structuring, modeling, development and deployment of artifacts and people" and its "*mission is to help organizations make better use of existing human, information and computer-based resources in order to build up the organization's knowledge and intelligence in a sustainable fashion*" [3].

ODE describes organizations as complex adaptive systems whose components are networks of people, processes, machines and other organizations. This has numerous implications:

1. The first implication is that a clear assumption is made that it is possible to apply principles of decomposition to the organization (system classification);
2. The second implication is that it is assumed that these systems change and are self-managing (the adaptive classification); and
3. Other implications come from classifying enterprises as social and complex: this means that enterprises will have particular properties namely: (1) *System Properties* such as *scalability, flexibility, stability, accuracy, robustness*, among others, which may be selectively targeted and most of the time imply the favoring of certain aspects over others (tradeoffs); and (2) *Emergent Values* or *Soft Properties* that are related to categories of social systems and result from the human dimension inherent to any enterprise.

ODE talks about properties such as *trust, motivation, loyalty, dedication,* and others. The last matter of relevance about ODE positioning, and that cannot be directly inferred by its ontological position is that a clear assumption is made that it is possible to apply the rigour of engineering sciences to approach organizational problems such as design and change [4].

Without having the tools to do so, organizations have to adapt to external and internal changes almost on a daily basis. Hence, today's organizations – profit-making and not-for-profit – have to develop new capabilities for continuous sensing, learning and adapting to ever-changing environments [3].

The mission of ODE is to, based on ODE principles (see Figure 1), help organizations make better use of existing human, information and computer-based resources in order to build up the organization's knowledge and intelligence in a sustainable fashion [3]. In this environment "Intelligence" is the term usually referring to a general mental capability to reason, solve problems, think abstractly, understand new information, learn from past experience and adjust to new situations [3].

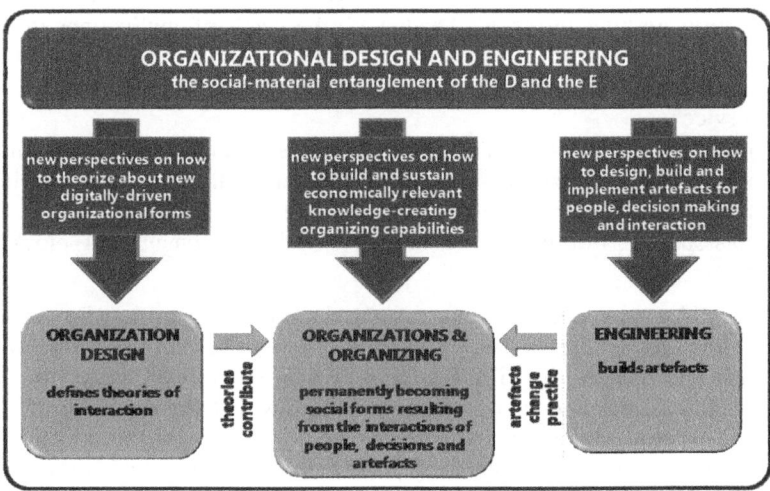

Fig. 1. Organizational Design and Engineering Entanglement of the D and the E [Source: [3]]

When applied to organizational settings, the representational technologies that Youngjin et al (2006) [5] talk about are crucial sources of self-awareness. Organizational self-awareness is part of the organization's knowledge creation process. It comprises a number of capabilities that jointly give the organization greater power to know itself. Self-awareness is of course a human capability, which can be greatly enhanced or diminished by the existing organizational designs combined with the existing computer-based artifacts [6]. The concept is obviously important because the larger the degree of self-awareness of any organization, the more cohesive it will become and the readier it will be at reacting to environmental change (external or internal). Organizational self-awareness and organizational agility go hand-in-hand.

The discipline of ODE is different from the traditional disciplines of Organizational Engineering because it tries to combine knowledge of social sciences with engineering sciences, enabling that the design of the social component of the organization (individuals, groups, values, culture, etc.) is combined to the rigour and the tools of the engineering disciplines [4].

Using the computer in the organization, as an essential tool to achieving strategy and managing own information and decision support elements, requires the determination of ISA, which consists of five sub-models [7].

- Enterprise Architecture (EA). Deals with aspects of the organization that are not directly related to the specific business and its operations, such as 'Mission', 'Vision', 'Strategy', and 'Organizational Goals';
- Business Architecture (BA). Relates with the materialization of the business strategy, implemented by business processes, representing the objects 'business process' and 'business purpose';
- Information Architecture (IA). Focus on what is that the organization needs to know to perform the operations, as defined in BA, characterized in EA and provides an abstraction of the organization's information needs, regardless of technology; it contemplates the objects 'feature', 'actor', 'observable state 'and 'activity';
- Application Architecture (AA). Deals with the needs of applications in data management and support of business, being independent of the software used to implement the different systems and includes the objects 'component SI' ('block IS') and 'service';
- Technological Architecture (TA). Handles all the technology behind the implementation of the applications as defined in the AA, as well as the necessary infrastructure for the production of support systems, business processes, and it considers concepts such as 'Information Technologies (IT) component', ('IT block') and 'IT service'.

One of the ISA components encompasses the definition of Business Processes, which can be defined as the "*set of interrelated and inter-performing activities that transform inputs into outputs*" [8]. The Business Process embody the generation of value-added businesses and is decomposed into activities that require a set of resources, human and material, in a defined time, which could contribute to their achievement.

The inability to run the processes, depending on its nature, leads to impairment of the success of an organization resulting in a set of incalculable damage that may even lead to loss of competitiveness and consequent market exit.

2.3 Enterprise Architecture and Enterprise Governance

"Architecture" when applied to the enterprise is denominated by Enterprise Architecture (EA). Within this context is also important to define what an enterprise is and what enterprise architecture is. Enterprise can be defined as *"any collection of organizations that has a common set of goals and/or a single bottom line"* [9].

While enterprise architecture provides a holistic view and captures the essentials of the business, and its evolution is very helpful in keeping the essentials of the business, so allowing for maximal flexibility and adaptability towards business success, good enterprise architecture provides the insight needed to balance these requirements and facilitates the translation from corporate strategy to daily operations [10].

A definition of EA can be *"a coherent set of descriptions, covering a regulations-oriented, design-oriented and patterns-oriented perspective on an enterprise, which provides indicators and controls that enable the informed governance of the enterprise's evolution and success"* [11].

Theories exist, today, to help to develop the Architecture. For instances, the Enterprise Ontology, can be defined *"as the realization and implementation independent essence of an enterprise, in short, as the deep structure behind its observable surface structure"* [12].

Enterprise Governance is a recent term that introduces the governance perspective that includes the strategic and operational performance with focus on compliance and performance concerns [13]. Enterprise governance is considered as *"the set of responsibilities and practices exercised by the board and executive management with the goal of providing strategic direction, ensuring that objectives are achieved, ascertaining that risks are managed appropriately and verifying that the organization's resources are used responsibly"* [13].

2.4 The Business Motivation Model (BMM)

The Business Motivation Model (BMM) [14] is a framework to develop the architecture of the Business in an organized way. The BMM has five major distinct areas: "Ends", "Means", "Influencers", "SWOT" and "Potential Impact". "Ends", which states what the organization wants to achieve, is composed by "Vision", "Goals" and "Objectives"; "Means", the way the organization uses to achieve its "Ends" includes the "Course of Action" and within this, "Strategy", "Tactics", "Business Policies " and "Business Rules"; Internal and external "Influencers" perform actions that could significantly impact both the "Ends" and the "Means". The "Strength, Weakness, Opportunities and Threats (SWOT) Analysis" lets one to know what impact these influencers have in the "Means" and in the "Ends"; "Potential Impact" can limit or jeopardize the activities of the organization.

Although it is a model revealing the behavior of the organization pursuing what it wants to be achieved, and how it will be accomplished, the BMM intends to motivate the components of the organization. Through this model, the elements understand the desired outcomes of the organization and how they are achieved, so there is a greater motivation on the part of its constituents.

"Mission" indicates the main activities of the organization, while "Vision" indicates the state that is sought and amplified by "Goals" and "Objectives".

"Course of Action" includes both "Strategy" and "Tactics". "Strategy" means the right approach to achieve the "Goals"; "Tactics", in turn, in relation to "Strategy", tend to fill in a shorter period of time, and have a more narrow perspective. "Tactics" are the tool for the achievement of defined "Objectives".

Directives, in any organization, serve to rule the "Course of Action". "Business Policies" are, in comparison with the "Business Rules", less structured, less discrete and not so small. On the other hand, "Business Rules" are highly structured, very thorough, presenting the standard vocabulary of the business, authorizing, restricting or guiding the work of the organization in specific areas.

The "External Influences" are those that stand outside the organization and create an impact on the application of "Means" or achievement of "Ends". The "Internal Influences" come from within the organization and have an impact on employment of "Means" and in the achievement of "Ends".

A SWOT analysis, according to the BMM, is a judgment on an Influencer, which affects the organization in its work to implement its "Means" or achieve its "Ends", that is, an analysis of strengths and weaknesses, opportunities and threats.

With the development of the SWOT analysis the "Potential Impact" can be anticipated, that is, to anticipate what impact the "Influencers" will have on "Means" or "Ends", positively or negatively. While negative influencers present a high Risk to the activity of the organization, positive influencers could, in turn, be used as a way to potential reward.

Risk arises from negative impacts indicating the probability of loss; naturally, without an analysis on the influencers, one cannot know the risk associated. The potential reward comes from positive results, indicating the winning probability. Like in the risk, in the absence of an analysis on "Influencers" the organization will not know what benefit could draw from them.

2.5 Self-Awareness, Flexibility, Agility and Change

There are numerous methodologies and methods related to the change of the organization which involves the establishment and articulation of the assumptions needed to start manufacturing processes and analyze and report how those really changed the organization.

Concepts related to the transformation and innovation, as self-awareness, flexibility and agility, are treated by various disciplines, including ODE, which experiments, analyzes and concludes about how concept application has effectively produced change in the organization.

This section describes and relates the principles of ODE to concepts applied to the transformation and innovation in the organization, placing them in the midst of change strategies advocated by modern scientists such as John Kotter (1996) [15].

In the context of change, it is also important to define a good strategy that encompasses solutions on the way to deal with resources in the process, including how to lead and avoid errors that may occur.

Concepts Associated with Transformation Processes and Innovation

In organizations in which there is the need to create deep changes, the concepts Transformation and Innovation are totally related [16].

Transformation means a continuous process, not having a specific end. Transformations are used to create or anticipate the future, raising ways of leading with the co-evolution of concepts, processes, organizations and technology. Change, in any of these areas, implies change of all [17].

Innovation means an introduction of something "new and uncommon", and it is the "heart" of any transformation [17]. It can also be defined as the development of new products, services, technology, work processes, markets and organizational structures [18].

Transformation and innovation increase, when certain important aspects of organization performance, such as self-awareness, flexibility and agility, are well succeeded.

Awareness consists of being conscious of the current moment, and it is important to the knowledge of how the world presents itself and to the adequacy of the actions to the reality [19].

The agility of an organization consists on its ability to move fast, decisive and effectively in anticipating, initiating and obtaining advantages from the change [16].

Flexibility is the capability of reaching success in different ways.

Change
Usually, change happens to adapt the organization to its strategy, and being the human resources the most valuable asset, they are also the ones that help to aid or hinder the process of change. The type of leadership can also make possible, help or difficult the happening of the change process. There are methods to initiate change and to determine problems that may arise in its implementation.

This section briefly discusses all the factors outlined above, describing the need for a strategy to implement change, to increase and demonstrate the feasibility of this complex process.

Strategy
Change, to be achieved, must be properly planned and balanced in order to be consistent. Thus, it is crucial that there is a strategy to implement change, to increase and demonstrate the feasibility of this complex process.

The Balanced Score Card is a Strategic Management methodology, composed of four different perspectives: financial, customer, internal processes, learning and growing [20][21].

Human Resources in Change
The success of the change processes in organizations depends on their human resources, which sometimes tends to create resistance to change. It is therefore essential to adopt a strategy for change that ensures that, by the human resources, acceptance, membership and involvement in change.

For this, it is important to know the best way to bring human resources to perform the desired actions. In this context, who is leading the change may take several options in what concerns his/her leadership:

- Education and communication;
- Participation and involvement;
- Slackening and support;
- Negotiation and agreement;
- Manipulation and cooptation;
- Explicit and implicit coercion.

According to the situation, the leader may choose one or more leadership styles mentioned above. They all have advantages and disadvantages, with consequences that must be carefully considered.

Driving Change

The changes need to be properly led to be successfully completed, so it is necessary to know how to properly target a transformation. According to John Kotter (1996) [15] there are eight processes which make possible to change in order to minimize the deviations from the transformation process implementation. These eight processes are:

- To establish a sense of urgency. It is essential to obtain the necessary cooperation and bring together a group with enough power and credibility to guide the effort or to convince key people to spend time to create and communicate a vision of change.
- To create a guiding coalition. In a process of change it is always necessary a strong guiding coalition, with the proper composition, levels of trust and common goals.
- To develop vision and strategy. Vision is constituted as an image of the future. In a process of change, an adequate vision allows to achieve three important aspects: first, to clarify the general direction of change, saying it is necessary to move in and out of the current situation; secondly, to motivate people to take action in the desired direction; and finally, to coordinate the actions of many people quickly and efficiently.
- To communicate the vision of change. The real power of vision is only triggered when the majority of people involved in the change have a common understanding of the objectives and direction. This can help to motivate and coordinate the types of actions that create the change.
- To empower broad-based action. This phase is to train employees about the actions that must occur.
- To generate short-term victories. Short-term victories are important because they make the change last. The statement "short-term victories" consolidates the actions and allows the identification of the "next step". Short-term victories must: i) be highly visible in order to allow a large number of people to find that the results are real, ii) be unambiguous and iii) clearly associated to change.
- To consolidate gains and produce more change. After reaching the objectives above, it is necessary to consolidate these gains. Moreover, it is profitable to organizations to continue to produce more change in order to profit and justify what has been done in the past. This is a step that can last for years, the more profound is the change, the more interdependent systems are changed.

− To anchor new approaches in the culture. After the desired change is finally implemented, it is necessary to anchor it in the organizational culture. Organization culture is a very powerful factor: individuals are selected and indoctrinated accordingly with the existing culture; it is exercised through hundreds or thousands of persons and most of the times become difficult to challenge or even discuss.

3 Flying

Ron Person (2008) in his book [22], states that leading a business in the current high-speed world is not that different from flying a high-speed jet. An objective metaphor was set by a conversation between a possible passenger, boarding an aircraft, and the pilot.

− "You: "What is our destination?""
− Pilot: "The crew got together and talked about a destination. We couldn't come to an exact agreement, so we decided to go somewhere out West. If something better comes up while we're en route, we might change direction."
− You: "What route will we be taking?" (Maybe I'll still go. It sounds adventurous, although it could be a waste of time and fuel. It shouldn't be too dangerous.)
− Pilot: "Well, we aren't sure about the exact route, but I've been that general direction before, so I don't need maps. I'm experienced."
− You: "I notice that your cockpit dashboard seems a bit sparse. There aren't any flight instruments — just stacks of paper. How will you monitor the flight?" (This is starting to sound a bit iffy. The pilot may be experienced, but how will he communicate his experience to the copilot, the flight engineer, the steward, the ground crew, other aircraft, and the Federal Aviation Administration?)
− Pilot: "Well, we're comfortable with the detail of printed reports. While we're in flight, I can request a short stack of printed reports that give me airspeed, altitude, attitude, and heading. The copilot gets a larger stack with operational data about radio settings, fuel, hydraulics, and technical details. We have to ask for the data, but it only takes a few minutes to get new reports — so we're in pretty good shape as long as everything stays stable and we don't have mechanical, weather, or crew problems.""

Flying is exactly the opposite of what is described in the metaphor. Flying is all about planning, become aware of every factor that affects not only the flight but also the flying environment.

Common errors, in the flying business, usually cost lives. Flying is all about planning, detail, awareness, precision, learning, controlling, analyzing and reporting in real time trying to forecast and mitigate exceptions that can result in error.

Comparing the standard organization to the flying organizations is an interesting exercise since most of the principles are common. The main difference is, perhaps, the cost of errors is immediately visible. Flying organizations have a very own culture that can be taken to the organizations in general.

Flying, itself, is a science of proven success. The several concepts associated with flying can also be taken to the organizations in a way that can create added value and situational awareness.

Thinking the organization as an aircraft can be surprising. Flying culture implies complying with strict rules on every component. Since this paper refers the operational and support components of an organization, a description of the Flying Culture effects will be described on both components.

This chapter presents several principles and concepts associated with the flying culture, on the architecture, personnel and material components.

3.1 Operational Crew Preparation

Flying implies a set of pre-requirements in several areas. If an actor (pilot, navigator, other) wants to occupy one flying position he or she has to follow a specific path comprehending four phases: i) health and psychological check; ii) environment courses; iii) aircraft training; iv) operational [23].

– Health and psychological check:
  – Health and body condition. First of all the candidate has to go a strict and very demanding set of tests. Extremely good eye vision and accuracy, spinal straighten is also mandatory.
  – Psychological. A good psychological condition is also essential. The candidate has to go through extensive testing and be present, at least once, on an interview with a psychologist.
– Environment courses:
  – Physiological course. A full physiological course is given where the candidate learns how to deal with problems that may arise within the air environment.
  – Water and land survival courses. Full water and land survival courses are given where the candidate learns how to deal with problems that may arise within the sea and land environment.
– Aircraft training:
  – Aircraft theoretical course. A very detailed and extensive aircraft course is given comprehending detailed information about essential systems, emergency procedures and emergency maneuvers.
  – Ground training. Comprehends aircraft systems training on the ground inside the cockpit. Emergency situations are taught and simulated to the possible extent.
  – Aircraft simulator. Simulator sessions are very intensive and are intended to simulate aircraft systems malfunction and the expected reaction from the crew.
  – Flight training. Trains the pilot on flying the aircraft and perform the basic flying procedures.
– Operational training:
  – Operational qualification. Provides operational qualification using aircraft capabilities.

Once the operational qualification is attained, the crew member regularly performs refresh courses and is submitted to semester testing, not only on the aircraft systems but also on the operational components. Pilots perform simulator training at least once a year.

3.2 Maintenance Crew

Aircraft Maintenance is the most important piece of support to the aeronautical business. Maintenance assures the aircrafts' continuing airworthiness, performing scheduled and non-schedule inspections and getting them ready and mission equipped to fly.

Maintenance personnel has to be trained to comply with high standards. Each person working on an airplane needs to have his or her technical formation, then needs to go on extensive courses about aircraft maintenance, and finally needs to obtain a qualification for performing a certain task on the airplane.

Every work is to be performed, in accordance with the published maintenance publications, by the worker and a supervisor and upon completeness, needs to be certified by an inspector granting that it has been done according to all the approved publications, norms, quality standards and inspection instructions.

3.3 Mission Planning

In the air there is little time for reasoning. Decisions must be made quickly and accurately; therefore, careful planning is essential to every flight. A smooth, successful mission requires a step-by-step plan which can be followed from take-off to landing. The main steps taken for mission planning are the following [24]:

- Select the destination and get all the information available;
- Select the route and map it;
- Flight Plan (time and resources);
- Fly;
- Stay on course, speed, altitude and planned fuel consumption;
- Check indicators permanently;
- Update en-route communications, navigation aids and alternate aerodromes;
- Debrief;

Due to its importance on the creation of self-awareness about all phases of flight, including en route and destination the next paragraphs describe with some detail the phases identified previously.

3.4 Destination Selection

A typical mission planning includes, after the destination is selected, consulting a extensive set of publications in order to get the necessary situational awareness of all the factors that can affect the mission en route and on the destination (Air Navigation, 2001):

- Flight information publications (FLIP). Complete aeronautical information concerning air traffic systems is published in FLIP.
- General planning (GP). This document is revised every 32 weeks with planning change notices (PCN) issued at the 16-week midpoint. Urgent change notices (UCN) are issued as required. The FLIP *GP* document contains information that is applicable worldwide. It is supplemented by the information published in seven Area Publication Sections.

- Area planning (AP). Located behind GP in the FLIP Planning document binder, contain planning and procedure information for a specific geographical area. The document is amplified by publications that contain a tabulation of all prohibited, restricted, danger, warning, and alert areas. In addition, they contain intensive student jet training areas, military training areas, and known parachute jumping areas within their specific geographical area.
- Planning change notices (PCN). These are in textual form and are used to update the FLIP.
- Flight information handbook (FIH). The FIH contains information for in-flight use. Sections include emergency procedures, national and international flight data and procedures, meteorological information, conversion tables and frequency pairings, standard time signals, and FLIP/NOTAM abbreviations and codes.
- FLIP En route charts. Charts portray airway systems, radio aids to navigation, airports, airspace divisions and other aeronautical data for IFR operations
- FLIP En route supplements (ERS). One is published for each geographical area. Each supplement contains an airport or facility directory, enroute procedures, special notices, and other textual data required to support FLIP *En Route Charts*. Airport sketch details include airport identification, city name, distance, direction, and elevation, as well as a diagram of each airport.
- Area and terminal area charts. These charts are large-scale graphics of selected terminal areas.
- Approach and departure procedures. Departure Procedures (DP) and terminal instrument approach procedures contain the approved departure and instrument approach procedures. Each instrument approach procedure shows an airport sketch, with additional data if necessary, for an approach under IFR conditions.
- Terminal change notices (TCN). TCN contains revisions to approach procedures and are published normally at the midpoint of the FLIP terminal booklets.
- Standard terminal arrival route (STAR). STAR contain preplanned IFR air traffic control arrival routes and are published in graphic and/or textual form. STAR provide transition from the en route structure to a fix or point from which an approach can be made.
- Notice to airman (NOTAM). A NOTAM is a message requiring expeditious and wide dissemination by telecommunication means. NOTAM provides information that is essential to all personnel concerned with flight operations. NOTAM information is normally in the form of abbreviations or a NOTAM code. The *FIH* contains an alphabetical list of these abbreviations.

Upon destination selection, all the information about the destination should be retrieved. Knowing all the information concerning the destination is essential to route selection.

3.5 Route Selection and Mapping

When planning a route to be flown, many factors enter into consideration. The route may be dictated by operational requirements of the mission; it may be a preplanned route, or the navigator may have the prerogative of selecting the route to be flown. In any case, definite factors affect route selection and the navigator must be aware of them.

In most cases, a direct route is usually best because it saves both time and fuel. Such things as airways, routing, high terrain, and bad weather, however, can affect this. The direction of prevailing winds can affect route selection because the proper use of a jet stream often decreases total flying time, even though a direct route is not flown.

Once a route is established, navigation charts appropriated to the intended flight path should be selected. Correct selection depends on distance to be flown, airspeeds, methods of navigation, and chart accuracy.

A great circle is the shortest distance between two points. One can save considerable distance by flying a great circle course, particularly on long-range missions in polar latitudes. A straight line on a gnomonic chart represents a great circle course. One way to flight plan a great circle course is to plot the entire route on a minor detail chart, and then transfer coordinates to charts more suitable for navigation. Select coordinates at intervals of approximately 300 nautical miles (NM). Once the route is plotted, record true courses and distances for each leg of the mission on the flight plan.

The method of navigation is determined by mission requirements and the flight mission area. Select charts for the mission which are best suited to the navigational techniques chosen (for example, radar missions require charts with representative terrain and cultural returns for precision fixing and grid missions require charts with a grid overlay). Once the route is selected and drawn the following information has to be retrieved:

- Alternate Airfield. An alternate airfield is where an aircraft intends to land if weather conditions prevent landing at a scheduled destination. Occasionally, an airfield may also be identified as an alternate for takeoff purposes. This procedure is at the direction of a major command that authorizes the use of lower minimums for takeoff than for landing.
- Emergency Airfields. During flight planning, select certain airfields along the planned flight route as possible emergency landing areas and then annotate these airfields on the charts for quick reference. Consider the following factors when selecting an emergency airfield: type of aircraft, weather conditions, runway length, runway weight-bearing capacity, runway lighting, and radio NAVAIDs. The NOTAMs for these airfields should be checked prior to flight.
- Highest Obstruction. After the route has been determined, the navigator should study the area surrounding the planned route and annotate the highest obstruction (terrain or cultural). The distance within which the highest obstruction will be annotated is IAW governing or local directives. The highest obstruction will be taken into consideration when determining the minimum en route altitude (MEA) and in emergency procedures discussion.
- Special Use Airspace. When determining the flight plan route, the locations of special use airspace will have to be considered. The best way to find the locations of the areas is by checking an en route chart. After the route is determined, any special use airspace that may be close enough to the route of flight to cause concern (as per governing directives) should be annotated on the chart with pertinent information. Annotate time and days of operation, effective altitudes, and any restriction applicable to that area. These areas, when annotated on the chart, will assist the navigator with in-flight mission changes and prevent planning a route of flight that cannot be flown.

4   Mission: Changing the Air Force

Based on a set of measures approved by the Air Force Chief-of-Staff, framed in his Vision, and in order to improve the relationship between the Organization's strategy with its information systems, it has been developed an action plan with three phases that that have started in March 2009 [25].

The first phase, using concepts depicted in Sections 2.1 to 2.5 (the initiative was launched following the rules depicted in Section 2.5), intended to determine the Organization AS IS by performing the actions identified below. The second stage intends to consolidate all the activities and lay down the basis for the TO BE planning (stage 3).

This plan of action, still ongoing, was intended to be a catalyst for change by identifying several areas of action, as defined below:

- Development of cross-organization doctrine (concepts, procedures), establishing a building of publications for operation and maintenance;
- Modeling of Processes and Activities of the maintenance and operation;
- Establishment of metrics and indicators for decision support in the information systems;
- Standardization of repositories of information (operational, maintenance, personnel);
- Integration of the articles to record activity of operational and maintenance components;
- Control mechanisms establishment.

Before the start of the process, Air Force has developed/updated business rules that allowed the framework of information systems in:

- Mission of the Air Units to reflect the changes expressed in the Strategic Concept of National Defense and the Military Strategic Concept, aligning the elements of the mission with NATO doctrine and defining the mechanisms required to obtain indicators related to air activity [26].
- The definition of dynamic and flexible mechanisms that would keep the amount of personnel needed for the operation and maintenance of weapons systems [27] for the purpose of:
  - Automatically calculating, by the given the assumptions, the amount of personnel (operation and maintenance) for the various Air Units.
  - Assessing the existing information on weapons systems and, if necessary, the methodology and verification of information integration, correcting methods and procedures.
  - Quantifying variations caused by specific features of the Air Units.
  - Comparing the existing workforce to the current regime of effort of the staff planned.
  - Defining and establishing appropriate planning and management planning, integrating information from different systems prevailing in each fleet, within the information system of the Air Force.

– The legal publications for operation and maintenance, constitute the building of Air Force operational publications, defining the responsible entities for their development and update. It underpins all the actions proposed by the plan in order to implement the change in the doctrine.

The doctrine, military term similar to business rules, based on experience, best practices and lessons learned [28], defines how it is expected to employ existing capabilities in a particular operation. The military doctrine is also the basis for future thought, integrating new technology and new capabilities [28]. The building of operating and maintenance publications, creating a doctrinal framework, aims to standardize and align it with the strategy across the Organization.

The approach to the process organization represents a new paradigm. While the doctrine says how to do, the survey process says exactly what it is done in reality.

The creation of metrics and mechanisms for decision support presupposes the existence and definition of goals and objectives enabling the monitoring and acquisition.

The standardization of repositories and integration of articles of information register forms are elements of dematerialization of essential tools for knowledge of the Organization itself.

The creation of control mechanisms allows the definition of the control points necessary to achieve the objectives.

To make the recommended actions operational it was created an implementation plan, spread over thirteen months, and to create organizational knowledge there were two different repositories for exchanging ideas: a directory on the internal network for staff directly involved and a forum in Internet that allows global access to information needed for the discussion.

The change also includes other acts such as the spread of organizational engineering at the Academy and at the various Promotion Courses.

Up to now fifty business policy and business rules manuals were produced, 470 business processes and key performance indicators were identified and 12 master thesis were developed (six are already completed).

In addition, the Air Force improved substantially its self-awareness. Chapter 5 presents some of the artifacts created and the associated concepts.

5 Flying the Organization or the Organization as an Airplane

All the flying vocabulary can be applied to the organization and enterprise concepts with minor adjustment. In fact, when looking at several concepts they seem alike and they seem to elaborate on the GST principles shown in Section 2.1. Table 1 shows some of the concepts and their counterparts.

Crew concepts include crew identification, which explains what they do, how they do it, and how they behave in respect to the outside world. In organizations that can be shown by a business model.

The next paragraphs discuss the enterprise vocabulary and its application to the Air Force.

Table 1. Concepts using Enterprise Vocabulary versus Flying Vocabulary

Enterprise Vocabulary	Flying Vocabulary	Mission Planning (Section 3.3)
Mission	Mission	Mission
Vision	Destination	Select Destination
Goals	Route	Select Route and
Objectives	Waypoints	map it
Strategy	Flight Plan	
Tactics	Flight Level, Speed	
Business Policy	Flight Policy	
Business Rules	Flight Rules	
Internal Influencers	Crew, aircraft status	
External Influencers	Weather, Airports, Flight Condition	Flight Planning
	Route Choosing, alternates, com-	
SWOT	munications, country authoriza-	
	tions, equipment	
Potential Impact	Capability to fly and attain destina-tion	
Business Processes	Aircraft Processes	Fly
Performance Indicators	Instruments	On course on time
Performance Dashboard	Cockpit indicators	Check & Update Debrief

5.1 Crew Identification or Who Are We?

One very important aspect of organization is that every human member understands what he or she is contributing with his or her individual work.

A survey was done by the Portuguese Air Force Academy students in which they enquired a selected set of people about if they knew what the contribution was of the work that they were developing to a greater achievement. In general, the selected people, working in the strategic and operational level knew what they were working for in terms of strategic objectives. However, when asked about if they knew everything that the Air Force was doing, the answer was negative.

Answering questions such as: "Who are we?", "What do we do?", "What are our values?" is done by the Business Model [29, 30, 31, 32]. Therefore, it became important to complement the Business Architecture Concepts by adding information about the organization itself. The Air Force Business Model development had to follow specific requirements:

1. To appeal to patriotism, given the highly patriotic nature of armed forces;
2. To be made into an easy and readable symbol, an image easy to understand;
3. To be able to represent any level of the Organization;
4. To represent the Organizational Structure;
5. To show the corporate values and mission;
6. To reflect areas (local) of employment;

Figure 2 shows the generic model for the PoAF Business Model [33, 34].

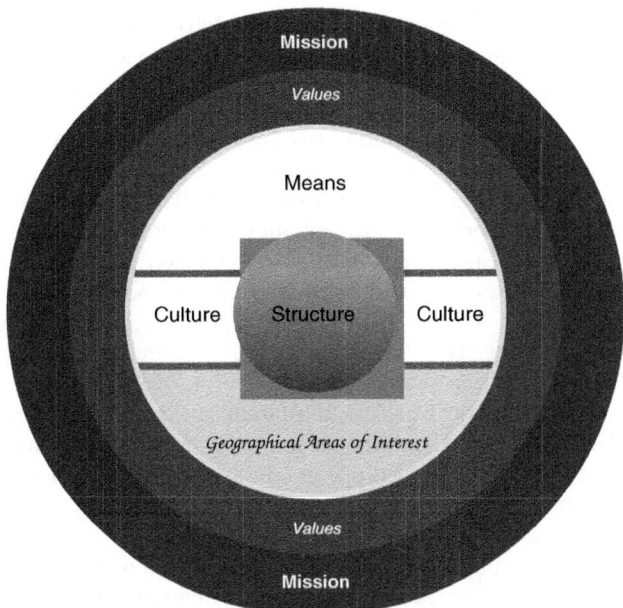

Fig. 2. POAF's Generic Business Model

While developing the Business Model for the Air Force, the necessary articles were included to meet the above requirements: the description of the operational means, actions, values, organizational structure and the sites where it operates. Figure 3 shows the model for the Air Force itself. The detail is described below.

The Portuguese Air Force is a branch of the Armed Forces that, in the operational area, operates different weapon systems that can be characterized by high specialization, such as, for example, speed, mobility, range and flexibility of employment in any type of theater. Integrating the system of national forces, the Air Force's Mission, among others, cooperates in an integrated way, in defense of the Republic, through execution of air operations and air defense of the national space [35].

The structure of the Organization consists of the following: at the strategic level of the hierarchy is the Air Force Chief of Staff (CEMFA), which is supported by the Air Force Staff (EMFA), the Inspector-General of the Air Force (IGFA), the Directorate of Finance (DFFA), by the Culture Organs (ONC) and by the Council Organs. These are followed by the operational level composed by four functional commands, the Logistics Command (CLAFA), the Air Command (CA), the Personnel Command of the Air Force (CPESFA) and the Education and Training Command (CIFFA). At the tactical level stand the Air Bases and the Air Units that operate the weapon systems.

In its normal activity the Air Force relates to the various entities including:

- The Portuguese Government, a regulatory element that ensures also a financial component;
- Other State Organs, such as the Presidency and the various Ministries and the Regional Governments;
- The military, for example, the General Staff of the Armed Forces, Army, Navy;

- The Portuguese Official Language's African Countries (PALOP);
- International Organizations such as the European Union and NATO;
- Other entities such as Universities, Media, Hospitals and Cultural organisms.

As part of its Mission, the Organization provides a range of services to the entities described above, being the most relevant: the defense of national airspace, air transport operations, patrol, search and rescue, maritime surveillance and medical evacuation, education (university and professional formation), research and development, health, courses of command and leadership and usage of the wind tunnel for aerodynamic tests.

The Air Force is a military institution that practices noble values, such as the following: "Do well to well serve", "Ethics of Rigor", "Responsibility", "Demand", "Culture of Merit", "Integrity", "Dedication", "Competence", "Justice", "Permanent availability", "Honesty", "Leadership" and "Discipline" [36].

In carrying out its specific Mission, the Organization operates around the Globe especially in Portugal, Main Land and Islands, the area of influence of NATO, the European Union and the Western European Union and also within the Community of Portuguese Speaking Countries, having recently participated in several national and international operations of which stands out Afghanistan.

In a clear reference to patriotism and the highest values of the Nation, the colors of the National Flag can be observed in the two outer circles and in the small inner circle.

In the center there is the organizational structure of the Air Force and various images. On left there are shown the students of the Air Force Academy, representing the essential training for any organization and on the right the F-16 fighter aircraft personnel, representing the operational field.

Also in the center, at the bottom, there are the local activities of that organization, at the top are the operational means of the Air Force, in a setting that highlights the fact that the Air Force flies in all ways and directions.

In the green circle are listed the values of all the Portuguese Air Force and in the red circle there are the various actions that the Air Force plays in civil and military components, in strict compliance with its Mission.

The Business Model enhanced the personnel situation awareness towards the Organization. One of the visible "things" is the broad scope of services provided and the relation with a multitude of external entities.

5.2 Concept Application

The following lines describe the Concepts identified in Table 1. As the change process is still happening, the Organization is still working on some of the areas. Nevertheless, a note has to be made to the fact that most of the described concepts are resulting from change and the shared effort between organization components with the aim of attaining goals and objectives while maintaining situational awareness.

Mission and Vision (select destination) or What do We Want?
The Air Force Mission is published by the Government on the Law of the Air Force Organization [35].

The destination was set with the Vision published by the General Chief-of-Staff (GCOS): *"In the multi-faceted coverage of the Mission, I envision an Air Force with a highly deployable nature, while maintaining a high degree of interoperability with*

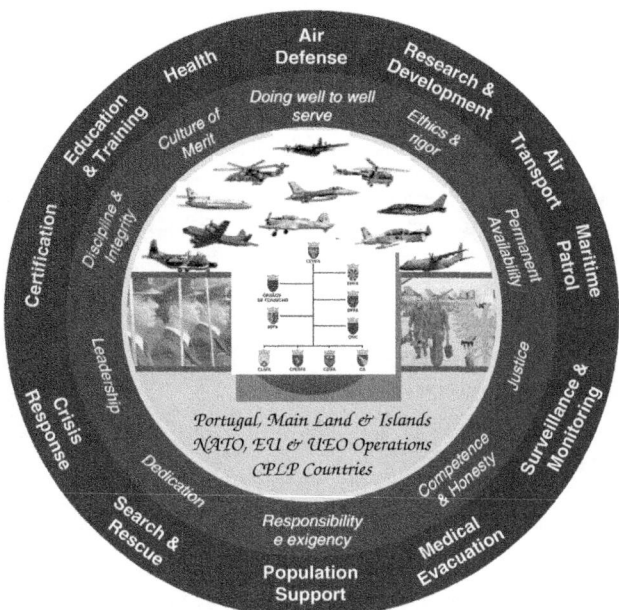

Fig. 3. Graphic representation of the Business Model for the Portuguese Air Force [33, 34]

other national and multinational forces, supported by the use of equipment that incorporate new technologies, served by a deployable command and control that enables operation in different environments, and a streamlined logistics, based on a modular structure, that eases expedited activation process. " [37].

It is important to mention that flying concepts depicted in Sections 3.3 to 3.5, when put together with concepts like the ones described in sections 2.1 to 2.5, provide a very useful insight in allying discipline of operating airplanes with scientific artifacts.

Naturally, before selecting the Vision (destination) a good analysis has to be performed (like the ones in mission planning, destination selection and route selection and mapping).

Goals & Objectives (Route & waypoints)

A model (shown in Figure 4) for aligning goals and objectives is being developed. The model was used to the definition of the management objectives for the year of 2010.

The model takes in consideration goals and the corresponding objectives. Priorities are a very important issue since the Organization relies heavily on external stakeholders for financial aspects that are crucial for objective definition.

Since the government funding scheme entails funding at several moments during the fiscal year, the Organization setup requires multiple configurations in order to be able to react to budget cuts by identifying impacts on the goals and corresponding objectives.

The model considers Goals (and priorities), key factors (number of hours flown) that will be affected by attaining the objective, organizational components (that will have to adequate own objectives) and business procceses (that attain objectives) which have to be accounted for while defining objectives.

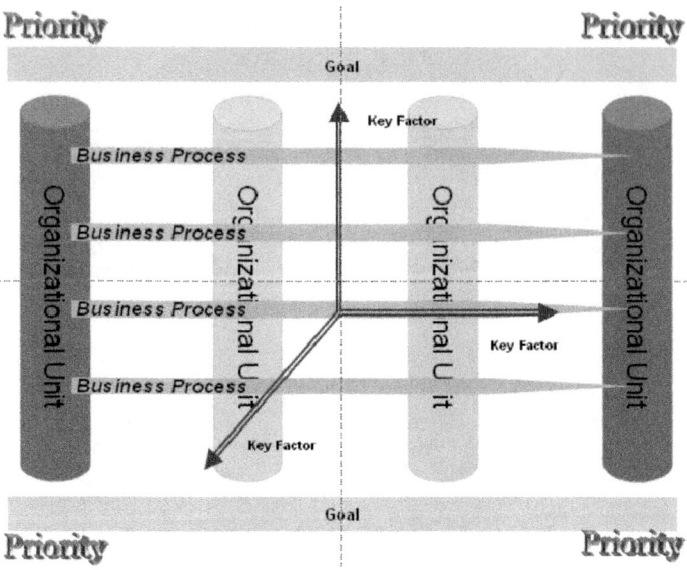

Fig. 4. Designing Objectives Model

An example can be shown.

- Goal: improving crew personnel draft in the next three years;
- Objective: improve drafting for navigators and pilots;
- Key factors: number of flying hours has to be increased;
- Business processes envolved: operational, training, image management (only to mention a few);
- Organizational components or units objectives (acronym description in section 5.3): EMFA, defining the number of personnel to be drafted; CIFFA, plan draft and courses; CA, fly demo sorties on specific areas; Public Relations, define a marketing campaign, DFFA, calculates the budget, etc...

By using the Model the Air Force makes sure that there is consistency betwen the several objectives in the Organization and also that, if a new configuration enters in place, due to financional constraints, what will be the objective to eliminate or reduce and what the financial impact will be.

Strategy & Tactics

A non-profit organization exists primarily to bring about changes in individuals and in society, and there is not the figure of "profit". Typically, these organizations exist to perform righteous or moral acts or causes to serve. However, like in profitable organizations, they should optimize their resources in order to add efficiency and effectiveness to their processes.

Strategy implementation in this type of organization differs from the others. While the Strategy in a profitable organization leads to produce the maximum profit as a result of its operation, Strategy in the nonprofit organization is merely a means of maximizing resources.

Currently the Air Force is developing a strategy map that can adequate to the organization's specifics and yet, represent the best way to achieve Goals and Objectives.

Business Policy & Business Rules
Business Policy and Business Rules are key orientations to strategy execution since they can define rules that restrict or expand the organization acts. In addition, they are the basis for development of the several configurations that drive adaptability in the Organization improving self-awareness and reaction times and easing the communications with external entities.

In the military, business policy and business rules tend to take either the form of directives issued by different levels or the form of doctrine.

On this subject, a hierarchy of manuals directive (example in Figure 5) was approved by the GCOS that intends to build doctrine at the different levels in a consistent manner [38].

As an example, the strategic level developed a concept of operations for each weapon system, which led to the development of two consistent manuals at the operational level: the concept of operational employment from the Air Command, and the concept of logistics support to operations from the Logistics Command.

At the tactical level, following the same principles, a set of rules and policies are also being developed.

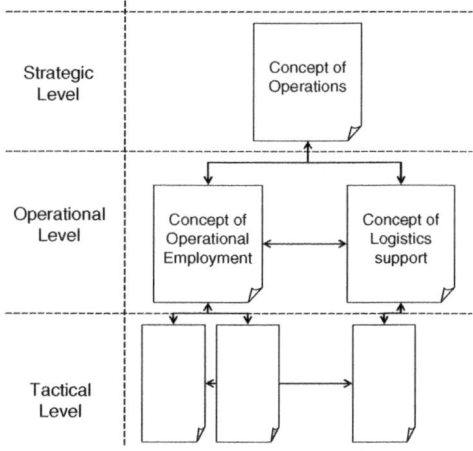

Fig. 5. Business Policy and Business Rules hierarchy

Internal & External Influencers, SWOT and Potential Impact
Internal and external influencers are directly connected to everything that the organization does. Externally influencers can produce enterprise governance artifacts and other type of negative and positive constraints. One example of an external influencer that produces governance on the Air Force is the Portuguese Government, which is also responsible for all the funding. The Portuguese Government also plays a customer role because it requests missions from the Organization.

Internal influencers act on all the definition of the organization artifacts. They are resources used by business processes; they can also act as internal governance creators in the form of business policy and business rules; they are also directly linked to goals, objectives, strategy formulation and to the SWOT and potential impact analysis.

The work created by the Air Force, as previously said, entails different configurations in order to allow situational awareness and rapid reaction times.

The research done by the Air Force Academy for the study and development of a strategy map considers, for each strategic action, the associated goals to fulfill, the corresponding business policies and rules, the related processes and the SWOT and Potential Impact analysis.

Business Processes
Since the business process is used to attain business objectives, Process Architecture can be understood as structuring the processes in management lines and value chain defining the required levels. Therefore, it is essential to find if the business processes attain the corresponding business objectives.

With this in mind, since the Air Force is identifying its business processes, a master thesis was conducted to establish the link to business objectives. The resulting work is now being used as a method for adding consistency between the two concepts. Figure 6 shows a picture of the scope [39, 40].

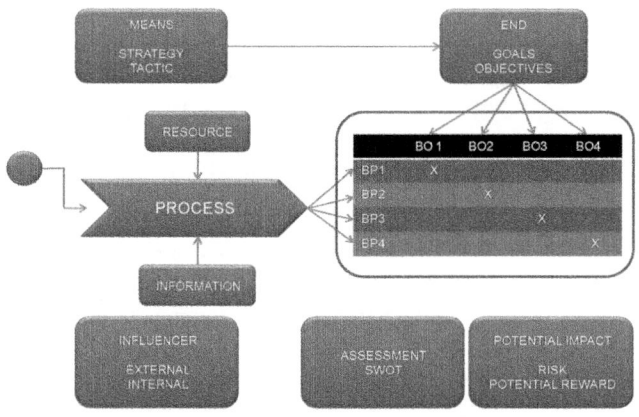

Fig. 6. Business Policy and Business Rules hierarchy [39, 40]

Performance Indicators & Dashboard
The importance of controlling parameters is vital to the functioning of any organization. Its importance becomes exponentially crucial when addressing large and complex organizations. Expertise on the analytical use of indicators and dashboards of the Portuguese Air Force was developed in another master thesis [41].

However, like in an airplane, to know what to control is vital. The work done was made around the Organization Key Factors (see section 5.2.2) in order to assist in the control, management and decision making process in a way benefiting organizational self-awareness and rapid reaction times. Using ODE principles and theories, basic artifacts were created. Examples are the missions performed by each weapon system, the number of people associated and the reification of the Organization Key Factors into tangible values.

Figure 7 shows a part of the organizational cockpit created showing indicators for different organization levels.

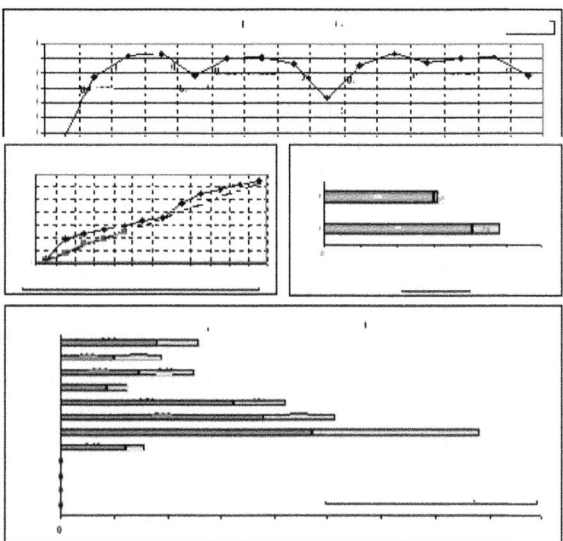

Fig. 7. A sample of the Air Force Organizational Cockpit

The formulation and definition of key factors brought along the need to measure efficiency and effectiveness that led to the creation of one number concept. Global effectiveness is given by one number and a scale of graduation. If needed, the concept provides the ability to drill down into the hierarchy and examine individual values.

5.3 Structuring the Air Force

Structuring well the organization in order to attain the goals is an essential step to success. In every organization, people should know what they have to do and therefore is essential to have track of people's functions, abilities and competences.

In addition, on the path to the business process oriented organization, it is very important to be able to compare what the people have to do, in the organization publications with what people actually do, in the business processes.

An analysis of the existing information in the organization's manuals and the way it was organized revealed some problems:

- Inexistent agreed semantics. There is no formal agreed terminology. However, tacit knowledge across the organization accepts that there are organizational entities (that have competencies) and job positions (that have functions and qualifications).

- Low consistency. Inexistence of horizontal and vertical verification of activities and competences (the first level organizational entity can have a competency that is not on the immediate level) with repetitions with a different text (for example: "participate in working groups" and "integrate working groups" or "produce reports" and "write reports")
- Regulations on paper. Paper regulations are very heavy, with a lot of sheets, little flexibility, hard to read, expensive due to the heavy spending of ink and paper, with high economical and environmental impacts.
- Induced inefficiency. Paper distribution limits the desired dematerialization, essential to smooth flow and process facilitation.
- Slow access to documents. Access to regulations is difficult and slow.
- Metrics inexistence. Inability to know the working hours associated to Job positions.
- Organization based on people. Hard maintenance based on people instead of Job positions.
- Difficult and expensive upgrades. Organization changes imply changing manuals, printing and distributing numerous copies.

Although the Air Force has its Information Technology (IT) System with some information about people that are occupying a certain Position, the EMFA, itself, has no electronic database that can act as a repository of information allowing for rapid query of some important issues like:

- What are the agreed upon semantics for the Organization? (Inexistent);
- What is the representation of the Organization? (Drawn on paper);
- Who works for whom? (Available on paper, does not answer to matrix groups that are created inside the organization with specific, time-limited tasks);
- What are the requirements, essential and desirable needed to occupy a job position? (Information exists on paper, however it lacks consistency between entities);
- What is the relevant information that one has to know about each job position (like, telephone number, job code, hierarchal dependency)? (Inexistent);
- What are the competencies needed to occupy a job position? (Information exists on paper, however it lacks consistency between entities);
- What are the functions of each job position? (Information exists on paper, however it lacks consistency between entities);
- What are the subjects that each job position deals with? (Available on the job position only with a very limited sentence);
- Who deals with a specific subject? (Available, but time consuming).

Developing an ontology, based on existing ontologies and theories [12] seemed to be a correct approach to establish an electronic repository of information accessible to all, while providing answers to the previous questions.

The concept has come to reality. It was developed in three phases and is one of the key items to be inserted in the new IS structure.

The idea behind is to apply to organizational positions the same concept that is currently applied to aircraft positions. If people cannot fly an airplane without going through the preparation depicted in Section 3.1, what would be the preparation necessary to "fly" an organizational position?

5.4 Shaping IT

IT shaping was one of the main objectives of the change process. An IT change plan [42] was prepared and the actions to identify the AS IS comprehended software acquisition to model business architecture and specification, information architecture.

The following phase will determine the TO BE and the design of a new architecture, using service-oriented architecture (SOA).

IT is also playing an essential role on personnel training and the setup of an internal directory and an Internet Forum where, upon registration, people could consult the "things" that have been done and provide feedback.

5.5 Back to the Basis or Assuring Knowledge

One of the essential aspects for maintaining knowledge is to involve the new generations. One of the key items in the Change process was to involve deeply the Academy and the Staff College in contributing to the development of concepts, theories, models and methods.

The Air Force Academy, with the Technical University of Lisboa, is contributing with the following master thesis:

- "A Business Model for the Air Force" [33, 34], completed;
- "Added Value Matrix representation for consistency between business objectives and business processes" [39, 40], completed;
- "Identification of Decision Support Indicators of Strategic Nature for the Portuguese Air Force" [41], completed;
- "Modeling the Effort Regime for the Air Force" [43, 44], completed;
- "Modeling the Flight Hour Cost for the Air Force" [45, 46], completed;
- "The Air Force Transformation Process" [47], completed;
- "A Strategy Map applied to a Military Organization", ongoing;
- "Two level (strategic and operational) objective definition", ongoing;
- "Modeling the Air Force Effort Regime Analysis Component", ongoing;
- "Business Rules alignment at different organizational levels", ongoing;
- "Changing the Air Force – Part II, conclusions" , ongoing;
- "Organizational and Design Engineering: Modeling the Flight Hour Cost for the Air Force using the Defence IS"; ongoing;

The Staff College contributed with five thesis on the promotion courses to senior officer.

6 Conclusion

The subject of maintaining self-awareness in an organization is discussed in this paper. How to draw, organize and manage an organization, in the human and material resources domains, considering i) multiple restrictions ii) critical needs of real time iii) various configurations is the key problem to be solved.

Concepts based on solid theories help the organization to stay self-aware and to know itself. Considering the organization to be a system, the general systems theory establishes a set of relations between two systems (or system and its exterior) that are expanded by other theories related to the organization like Organizational and Design Engineering, Enterprise Architecture and Enterprise Governance, the Business Motivation Model and self-awareness, agility and change.

To answer the questions above, the Portuguese Air Force Chief-of-Staff determined that the Air Force would conduct a holistic approach to the problem and a change process was initiated. The change process evolved three main items and had the objective of aligning IT with the strategy:

- Development of cross-organization doctrine (concepts, procedures), establishing a building of publications for operation and maintenance;
- Modeling of Processes and Activities of the maintenance and operation;
- Establishment of metrics and indicators for decision support in information systems.

Change included a strong support by the scientific community and comprehended master thesis development at the Air Force Academy.

Up to now fifty business policy and business rules manuals (at the different levels were produced, 470 business processes and key performance indicators were identified and 12 master thesis were developed (six are already complete).

As the discipline of flying deals with the same problems, the flying concepts were applied to the business. Personal preparation, mission, cockpit are being translated into their business counterparts.

As a result, the Air Force is identifying critical factors while studying and applying new concepts and resulting artifacts, such as:

- The business model (to enhance internal understanding);
- The business objectives selection model (to guarantee consistency between Vision, goals and business processes);
- The business rules hierarchy model (to hold consistency between the business rules at different organization levels);
- The business objectives/business processes consistency model (to hold consistency between the business objectives and the business processes);
- The organizational cockpit model (to verify strategy execution and that the organization is reaching the defined goals).
- The online organization (to verify what the organization's best structure is and to provide a set of best practices in organizational and personnel competencies).

Considering i) multiple restrictions, ii) critical needs of real time, iii) various configurations, entail a set of actions. Conclusions are:

- Since the Air Force is highly affected in its operation by external influencers, planning is an essential tool;
- Self-awareness comes together with the need of knowing exactly what the Vision (destination) is, and the corresponding goals and objectives (route and waypoints). Alternate selection is also a key point to attaining rapid response to environment changes;

- Strategy (route and waypoints), together with the business policies and business rules, play an essential role in how to achieve the objectives;
- Business processes and business objectives must be well aligned;
- IT plays an important role in Enterprise Architecture since it stands as the facilitating actor.

Taking in consideration what has been said before, one possible solution to attain operational effectiveness in a multiple restriction environment is to consider multiple configurations settings with a configuration manager.

Configuration entails a set of business concepts with different objectives and different rules.

The configuration manager holds the key to change configurations, in near real time, as a result of changes in the organizational world, caused either by internal or external influencers.

References

1. Maturana, H., Varela, V.: The Tree of Knowledge, Shambala, London (1992)
2. von Bertalanffy, L.: General Systems Theory. Main Currents in Modern Thought 71(75) (1955)
3. Magalhães, R., Rito, A.: A White paper on Organizational and Design Enginering, Center for Organizational and Design Engineering, INESC INOV, Lisboa, Portugal (2009)
4. Matos, M.G.: Organizational Engineering: An Overview of Current Perspectives, Master Thesis in Computer Science, IST/UTL, Lisboa (2007)
5. Youngjin, Y., Boland, R.J., Lyytinen, K.: From Organization Design to Organization Designing. Organization Science 17(2), 215–229 (2006)
6. Tribolet, J.M.: Organizações, Pessoas, Processos e Conhecimento: da Reificação do Ser Humano como Componente do Conhecimento à "Consciência de Si Organizacional" [Organizations, People, Processes and Knowledge: from the reification of the human being as a component of knowledge to the knowledge of the organizational self]. In: Amaral, L., Magalhães, R., Morais, C.C., Serrano, A., Zorrinho, C. (eds.) Sistemas de Informação Organizacionais. Sílabo Editora, Lisboa, Portugal (2005)
7. Abreu, M., José, T.: Considerações sobre a medição de factores soft nas organizações, Center for Organizational and Design Engineering, INESC INOV, Lisboa, Portugal (2008)
8. Instituto Português da Qualidade, ISO 9000:2000, Quality Management Systems – Fundamentals and Vocabulary, Lisboa (2000)
9. The Open Group: A Pocket Guide to TOGAF Version 8.1.1 Enterprise Edition. The Open Group (2007), http://www.opengroup.org/togaf
10. Lankhorst, M.: Enterprise Architecture at Work - Modelling, Communication and Analysis. Springer, Heidelberg (2005)
11. Op 't Land, M., Proper, E., Maarten, W., Cloo, J., Steghuis, C.: Enterprise Architecture – Creating Value by Informed Governance. Springer, Heidelberg (2009)
12. Dietz, J.: Enterprise Ontology: Theory and Methodology. Springer, Delft (2006)
13. Hoogervorst, J.A.P.: Enterprise Governance & Enterprise Engineering. Springer, The Netherlands (2009)
14. The Business Rules Group (2007), The Business Motivation Model: Business Governance in a Volatile World (consulted October 2009), http://www.businessrulesgroup.org/bmm.shtml

15. Kotter, J.P.: Leading Change. Harvard Business School Press, Boston (1996)
16. Agility Consulting: The Agile Model, http://www.agilityconsulting.com (consulted in January 2010)
17. Alberts, D.S., Hayes, R.E.: Campaigns of Experimentation: Pathways to Innovation and Transformation. CCRP Publications Series (2005)
18. Cunliffe, A.L.: Organization Theory. SAGE Publications Ltd., London (2008)
19. Nifco, N.: A Conceptualization of Knowledge Management Practices Through Knowledge, Awareness and Meaning (2005), http://www.ejkm.com/volume-3/v3i1/v3-i1-art5-nifco.pdf (consulted in October 2009)
20. Kaplan, R., Norton, D.: Strategy Maps – Converting intangible assets into tangible outcomes. Harvard Business School Press, Boston (2004)
21. Kaplan, R.S., Norton, D.P.: The Balanced Scorecard: Translating Strategy into Action. Harvard Business School Press, Boston (1996)
22. Person, R.: Balanced Scorecards and Operational Dashboards with Microsoft Excel. Wiley Publishing, Inc., Indianapolis (2009)
23. Headquarters: Flying Rules for Crews. Portuguese Air Force, Lisboa, Portugal (2010)
24. Secretary of the Air Force Office, Air Force Pamphlet 11-216 Air Navigation, Washington DC (2001)
25. Headquarters: Change Directive. Portuguese Air Force, Lisboa, Portugal (2009)
26. Headquarters: Mission for the Air Units Directive. Portuguese Air Force, Lisboa, Portugal (2007)
27. Headquarters: Operational and Maintenance Personnel for the Air Units. Portuguese Air Force, Lisboa, Portugal (2009)
28. Air Force Basic Doctrine: United States Air Force Doctrine Center, Washington DC (2003)
29. Jägers, H., Jansen, W., Steenbakkers, W.: New Business Models for the Knowledge Economy. Gower Publishing Limited, Hampshire (2007)
30. Ramos, R.: Conhecimento de TI – Modelo de Negócio e Estratégia. Você tem? (2009), http://www.conhecimentoeti.com/2009/09/modelo-de-negocio-e-estrategia-voce-tem.html (consulted October 2009)
31. Porter, M.E.: What is Strategy? Harvard Business School Publishing Corporation, Boston (1996)
32. Ulwick, A.W.: Business Strategy Formulation: Theory, Process, and the Intellectual Revolution. Quorum Books, Westport (1999)
33. Leal, P., Páscoa, C., Tribolet, J.: Organizational and Design Engineering: A Business Model for the Portuguese Air Force, Master Thesis, Department of Universitary Education, Portuguese Air Force Academy, Sintra, Portugal (2010)
34. Leal, P., Páscoa, C., Tribolet, J.: A Business Model for the Portuguese Air Force. Springer CCIS Series on Minutes of the CENTERIS 2010, Conference on Enterprise and Information Systems, Viana do Castelo, Portugal (2010) (forthcoming)
35. Law of the Portuguese Republic: Air Force Organization, Portuguese Government, Lisboa, Portugal (2009)
36. Araújo, L.: Speech of the Portuguese Air Force Chief-of-Staff (2008)
37. Headquarters: Planning Directive for 2008-2010. Portuguese Air Force, Lisboa, Portugal (2008)
38. Headquarters: Planning Directive for 2008-2010. Portuguese Air Force, Lisboa, Portugal (2008)

39. Belo, N., Páscoa, C., Tribolet, J.: Value Model for Enterprise and Process Architectures Alignment Verification, Master Thesis, Department of Universitary Education, Portuguese Air Force Academy, Sintra, Portugal (2010)

40. Belo, N., Páscoa, C., Tribolet, J.: Value Model for Enterprise and Process Architectures Alignment Verification. Springer CCIS Series on Minutes of the CENTERIS 2010, Conference on Enterprise and Information Systems, Viana do Castelo, Portugal (2010) (forthcoming)

41. Rodrigues, M., Páscoa, C., Tribolet, J.: Identification of Decision Support Indicators of Strategic Nature for the Portuguese Air Force, Master Thesis, Department of Universitary Education, Portuguese Air Force Academy, Sintra, Portugal (2010)

42. Headquarters: Planning Directive for the new generation of IS. Portuguese Air Force, Lisboa, Portugal (2009)

43. Alves, A., Páscoa, C., Tribolet, J.: Modeling the Effort Regime for the Air Force, Master Thesis, Department of Universitary Education, Portuguese Air Force Academy, Sintra, Portugal (2010)

44. Alves, A., Páscoa, C., Tribolet, J.: Modeling the Effort Regime for the Air Force. Springer CCIS Series on Minutes of the CENTERIS 2010, Conference on ENTERprise and Information Systems, Viana do Castelo, Portugal (2010) (forthcoming)

45. Soares, J., Páscoa, C., Tribolet, J.: Modeling the Flight Hour Cost for the Air Force, Master Thesis, Department of Universitary Education, Portuguese Air Force Academy, Sintra, Portugal (2010)

46. Soares, J., Páscoa, C., Tribolet, J.: Modeling the Flight Hour Cost for the Air Force. Springer CCIS Series on Minutes of the CENTERIS 2010, Conference on Enterprise and Information Systems, Viana do Castelo, Portugal (2010) (forthcoming)

47. Malico, J., Páscoa, C., Tribolet, J.: The Air Force Transformation Process, Master Thesis, Department of Universitary Education, Portuguese Air Force Academy, Sintra, Portugal (2010)

Architecture-Based IT Portfolio Valuation

Marc M. Lankhorst, Dick A.C. Quartel, and Maarten W.A. Steen

Novay, P.O. Box 589, 7500 AN Enschede, The Netherlands
{marc.lankhorst,dick.quartel,maarten.steen}@novay.nl

Abstract. This paper describes the ingredients of an integrated IT valuation method that uses architectural models as its backbone. First, it investigates the link between the organization's mission and vision and high-level strategy, such as its value center approach and operating model. These strategic choices determine the aspects that need to be taken into account when assessing the value of the IT portfolio. The resulting business requirements can be modeled in conjunction with the enterprise architecture of the organization. This provides concrete insights in the contribution of these elements to the business. KPIs are then associated with business requirements on the one hand and architecture elements on the other hand, and measurement of these KPIs determines the operational performance of the organization and its IT. The paper uses an implementation of Bedell's method as an illustration of this approach.

Keywords: IT portfolio management, requirements management, enterprise architecture, IT governance, IT investment management, IT value management.

1 Introduction

After almost half a century of IT developments, many large organizations face an unfavorable ratio between old (existing) IT and new IT. Because old IT systems tend to be monolithic, unwieldy and inflexible, organizations experience maintenance as difficult and modernization to meet new business demands as improbable. Some organizations spend up to 90% of their IT budget in 2009 on maintaining the existing IT landscape, leaving only 10% for innovation. If this trend of increasing budget requirements for existing IT is not reversed, then in the nearby future no budget at all will be available for new IT. In the worst case, innovation is squeezed out completely and budgets to spend on existing IT may become insufficient to perform crucial maintenance tasks.

By focusing on the value of IT instead of considering costs only, organizations can decide which IT really contributes to their business goals and make a well balanced division into budgets for maintenance, exploration, realization and phasing out. Traditionally, IT has often been regarded only as a cost center in business case calculations. Its less tangible benefits have often been more or less neglected in portfolio management decisions. Furthermore, in the past information systems tended to be relatively stand-alone, supporting a single business silo. This made it easier to attribute their costs and benefits.

F. Harmsen et al. (Eds.): PRET 2010, LNBIP 69, pp. 78–106, 2010.

IT systems and services are more and more interwoven with the business and may support many different activities, generate independent revenue streams, attract new business, et cetera. To provide insight into these effects, a valuation approach is needed that encompasses the coherence of the entire organization, its products and services, processes, applications, and infrastructure, i.e., the enterprise architecture. Our study of existing literature showed that some IT valuation approaches have close associations with enterprise architecture. However, the process of translating architectural benefits into value needs a complete understanding of the relationship between architecture and business benefits. Unfortunately, little research is currently addressing this. With this work on architecture-based IT valuation we aim to fill this gap.

This paper describes the ingredients of an integrated IT valuation method, which uses architectural models as its backbone. First, we must investigate the business requirements that result from the organization's mission and vision and from its high-level strategy, such as its value center approach and operating model, as outlined in the next sections. These strategic choices determine the aspects that need to be taken into account when assessing the value of the IT portfolio. The resulting business requirements can be modeled in conjunction with the enterprise architecture of the organization. This helps in realizing traceability between business requirements and IT artifacts, which is needed to perform a well-founded portfolio assessment, and it provides concrete insights in the contribution of these elements to the business. KPIs are then associated with business requirements on the one hand and architecture elements on the other hand, and measurement of these KPIs determines the operational performance of the organization and its IT.

Business requirements and enterprise architecture are the main inputs to calculate the 'value' of IT systems and projects. The importance of different criteria to assess this value depends on the strategic direction of the organization. Strategic choices are linked to one or more business goals from which valuation criteria and performance indicators are derived. Depending on these criteria, different valuation techniques may be selected to analyze IT with respect to these criteria. Separate IT budgets may be allocated to limit the IT investments for each value center.

This paper is structured as follows. First, in Section 2 we introduce portfolio management as a way to control IT investments in an integrated way. Naturally, IT strategy and investments must be in line with the organization's strategy as a whole; this is addressed in Section 3, which describes the value center approach to IT strategy. Sections 4 and 5 then describe the implementation of the IT strategy in terms of the organization's operating model and the role of enterprise architecture in defining this. Section 6 describes how business goals and requirements can be made more concrete and related to the enterprise architecture, in order to realize the desired architecture. In Section 7, we describe Bedell's method [1] for assessing the contribution of IT assets and projects to these business goals, in order to build a balanced portfolio that provides an optimal allocation of investments. A tool implementation and example of this method is shown in Section 8. Sections 9 provides more detail on assessment criteria that may be used in this method, including quality attributes such as the well-known '-ilities' from software engineering and risk analysis criteria. Section 10 describes so-called valuation profiles, which link the IT strategy described in Section 3 to relevant assessment criteria. Section 11 provides an overview of related work, and Section 12 presents our conclusions and ideas about future work.

2   Portfolio Management

Before we can discuss IT valuation itself, we have to describe its context. The need for IT valuation usually arises in larger organizations when they recognize the need to manage their investments in IT as a portfolio, much like the way in which other (financial) investments would be managed. By managing investments as a portfolio, organizations hope to make better informed decisions, and achieve better overall outcomes, rather than considering projects on a case-by-case basis using subjective arguments. IT portfolio management is about managing and balancing the investments in IT to optimize their benefits in relation to their costs and risks [2]. An important element of portfolio management therefore is the valuation of IT projects and assets in terms of their costs, benefits, risks and contribution to strategic objectives.

Two kinds of portfolio can be considered in IT: the project portfolio and the application portfolio. The former considers and manages investments in future IT capabilities, whereas the latter considers and manages the value of existing IT assets. These two perspectives can be related as follows. Based on the valuation of the current portfolio of IT assets, various tactical decisions may be made to better align them with business goals. Examples of such decisions are to continue maintenance, to improve quality, to extend functionality, or to replace and phase out. Each of these decisions will result in the proposal of a project or program, which needs to be evaluated on its value and fit within the project portfolio. Once completed the projects deliver a new IT landscape, representing the new application portfolio.

Following [3], we consider the following three phases in the portfolio management process: strategic planning, individual project evaluation, and portfolio selection. The first phase comprises strategy development, methodology selection, determination of strategic focus and overall budget and resource allocation policies. The actual selection takes place in the second and third phase. In the second phase individual projects and programs are evaluated independently of other projects, but using common criteria and indicators. The third phase then deals with the alignment and balancing of the entire portfolio. In this phase projects are prioritized based on their strategic fit and other criteria including their interactions with other projects through resource constraints or other interdependencies. In accordance with current standards for portfolio management, such as Val IT [4], we add a fourth phase, portfolio monitoring, in which the portfolio's performance is measured and the portfolio is adjusted if necessary. The process is continuously repeated, although certain phases such as strategy planning and individual project evaluation may be skipped if they are not needed.

In the following sections we propose ways of dealing with each of these phases except the last. Sections 3 to 6 and 10 provide methods and techniques to support strategy planning. Sections 7 to 9 deal with methods to support both project evaluation and portfolio selection.

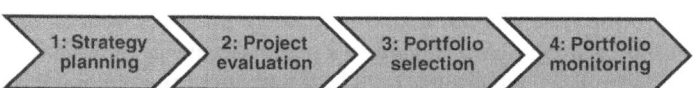

Fig. 1. Portfolio management process

3 IT Strategy: Value Centers

In putting a value to IT systems, projects and investments, it is highly important to first have a clear insight in the strategic choices the organization has made with respect to its IT operations. Many organizations see IT mainly as a cost center; its possible contribution to the overall business strategy is often overlooked.

Venkatraman [5] was one of the first to look at IT strategy in a broader manner. This work presents an important and well-founded strategy framework that supports a differentiation in IT goals: the value center orientation for IT. The main idea is that each center represents a different way of extracting value from IT resources. Note that the centers are interdependent. Venkatraman considers four different value centers (see Figure 2). *'The cost center reflects an operational focus that minimizes risks with a predominant focus on operational excellence. Service center, while still minimizing risk, aims to create IT-enabled business capability to support current strategies. Investment center, on the other hand, has a longer-term focus and aims to create new IT-based business capabilities. Finally, profit center is designed to deliver IT services to the external marketplace to realize incremental revenue as well as gain valuable learning and experience to become a world-class IT organization'* [6].

So on the one hand, there are the cost center and service center approaches, focusing on current business strategies. On the other hand, there are investment center and profit center that aim at maximizing opportunities from IT resources and shaping future business strategies.

For each center, specific business goals and performance indicators can be defined. This approach with different IT strategies fits with the focus of the IT valuation method our applied research project is constructing. The business strategy and the matching value centers provide input for the choice of valuation and assessment criteria for the IT portfolio. Note, however, that our method is not dependent on this specific IT strategy framework; as explained in Section 10, the strategic choices are used to determine which combination of assessment criteria is to be used (a so-called 'valuation profile'). Venkatraman's approach merely serves as a good and well-founded example of such a framework.

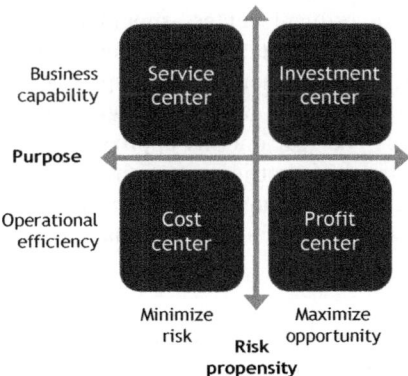

Fig. 2. The concept of a value center [5]

3.1 Cost Center

IT that is typically positioned as a cost center is not directly related to business goals. Examples are the operational infrastructure involving most data centers, telecommunications network and routine maintenance like installing and removing equipment, answering questions and administrative support. Specific performance metrics are used as decision criteria, which are not related to business metrics. Cost center works well when input and output can be clearly related, like doubling the budget results in a performance increase by factor 3. Relevant performance metrics are quantitative in nature, for example costs per unit of something, maintenance costs per unit, or costs per employee. Such measures need to be benchmarked against performance metrics of other organizations in order to be able to find opportunities for improvement.

3.2 Service Center

A service center aims to create IT-enabled business capabilities that drive current business strategy. IT resources create tangible current business advantages. IT is strongly related to business goals. Investment decisions are not solely based on costs but rather on improving service provisioning. Whether an IT system is a cost center or a service center depends on the organization. In this way, an IT system can be considered as a service center for the one organisation and a cost center for the other. For example service characteristics such as minimize downtime and improve reliability can also be considered as performance metrics. The main question in the service center category is whether an IT system gives the organization a competitive edge and differentiates the organisation from its competitors. So the purpose of use of an IT system is important and not the application and functionality in itself. From a service center perspective an organization should look at the degree that an IT system contributes to customer acquisition and retention.

3.3 Investment Center

An investment center has a future orientation. It focuses on innovations, for example creating new business capabilities by means of IT. This requires more than IT. New business capabilities are created with a unique combination of structure, processes, systems and expertise. Investment centers should focus on more than technology. Next to IT investments complementary investments will be needed to realise a business capability. That is, IT investments become part of a total package. Investment center involves resource allocations based on strategic redirection and reliance on IT for business innovations. The real options approach fits with the investment center rather than traditional financial metrics, since the real options approach takes risks and uncertainties into account. The investment center should be run as a venture capitalist. It requires the forward look of a business innovator.

3.4 Profit Center

A profit center has a focus on delivering IT products and services in an external marketplace. Next to financial benefits the intangible benefits should also be taken into account in investment decisions. The profit center needs an external, marketing orientation, instead of an internal captive monopoly. The profit center should work in value networks and partner with other companies in combining complementary skills and resources to deliver value.

4 Operating Model

Next to the commercial strategy that is chosen for IT operations, as outlined in the previous section, we also need to take into account the more operational aspects of the organization in defining an IT planning and valuation approach. As Ross et al. [7] show with numerous case studies, successful enterprises employ an 'operating model' with clear choices on the levels of integration and standardization of business processes across the enterprise (Figure 3):

1. *Diversification*: different business units are allowed to have their own business processes. Data is not integrated across the enterprise. Example: diversified conglomerates that operate in different markets, with different products.
2. *Replication*: business processes are standardized and replicated across the organization, but data is local and not integrated. Example: business units in separate countries, serving different customers but using the same centrally defined business processes. Example: a fast food chain replicating its way of working through all its local branches.
3. *Coordination*: data is shared and business processes are integrated across the enterprise, but not standardized. Example: a bank serving its clients by sharing customer and product data across the enterprise, but with local branches and advisers having autonomy in tailoring processes to their clients.
4. *Unification*: global integration and standardization across the enterprise. Example: the integrated operations and supply chain of a chemicals manufacturing company. This operating model should fit both its area of business and its development stage.

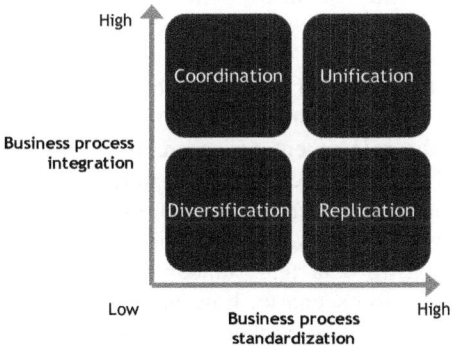

Fig. 3. Operating models [7]

5 The Role of Enterprise Architecture

Increasingly, the notion of architecture is applied with a broader scope than just in the technical and IT domains. The emerging discipline of enterprise engineering views enterprises as a whole as purposefully designed systems that can be adapted and re-designed in a systematic and controlled way. An 'enterprise' in this context can be defined as 'any collection of organizations that has a common set of goals and/or a single bottom line' [8]. Architecture at the level of an entire organization is commonly referred to as 'enterprise architecture', which can be defined as a coherent whole of principles, methods, and models that are used in the design and realization of an enterprise's organizational structure, business processes, information systems, and infrastructure [9].

Over the last years, the field of enterprise architecture has seen considerable developments. The toolbox of the enterprise architect nowadays comprises a wide array of methods, techniques and tools. The Open Group Architecture Framework (TOGAF) [8] has become the de facto standard way-of-working for architects, the ArchiMate design language [9, 10] is now an international standard for modeling and analyzing enterprise architectures, and a plethora of software tools implement these standards.

Only recently, hard figures are becoming available on the actual contribution of architecture [11]. A study by Slot, Dedene and Maes [12], based on an analysis of 49 projects, clearly shows the benefits of enterprise and project architecture. So having a good enterprise architecture practice may deliver direct and indirect cost savings and other benefits, because decisions are made in context: it offers a holistic view, showing the interdependencies between different parts of the enterprise.

5.1 Strategy Execution

Enterprise architecture captures the essentials of the business, IT and its evolution. The idea is that the essentials are much more stable than the specific solutions that are found for the problems currently at hand. Architecture is therefore helpful in guarding the essentials of the business, while still allowing for maximal flexibility and adaptivity. Without good architecture, it is difficult to achieve business success.

Architecture forms a strategic instrument in guiding an organization through a planned course of development. As we explained previously, an organization's 'operating model', as defined by Ross, Weill and Robertson [13], is highly important in defining its tactical and operational choices with respect to IT. This operating model should fit both its area of business and its stage of development.

Ross et al. explain the role of enterprise architecture as the organizing logic for business processes and IT infrastructure, which must reflect the integration and standardization requirements of the operating model. For example, ERP systems are used extensively by companies that have a unification strategy, since these systems are well-suited for both sharing data and standardizing business processes across the enterprise. In a diversification scenario, however, investing in an ERP system might be a wrong choice, since the varied collection of business processes and localized data do not lend themselves to the 'one size fits all' approach of such a system.

Next to this operating model, Ross et al. provide a stage model of the architectural development of organizations:

1. *Business silos*: every individual business unit has its own IT and does local optimization.
2. *Standardized technology*: a common set of infrastructure services is provided centrally and efficiently.

3. *Optimized core*: data and process standardization, as appropriate for the chosen operating model, are provided through shared business applications (e.g. ERP or CRM systems).
4. *Business modularity*: loosely coupled IT-enabled business process components are managed and reused, preserving global standards and enabling local differences at the same time.
5. *Dynamic venturing*: rapidly reconfigurable, self-contained modules are merged seamlessly and dynamically with those of business partners.

In practice, most companies are still in stages 1–3. Investment decisions should be guided by the chosen operating model and the current and desired stage of an organization. E.g. if an organization wants to move from stage 1 to stage 2, the focus should be on standardizing and centralizing IT infrastructure in order to achieve efficient operations. The contribution of IT systems and projects to achieving the desired stage, in concordance with the operating model, should be a core criterion in valuating these systems or projects.

Another reason for using enterprise architecture in investment decisions is that it provides a coherent view of the various dependencies between IT systems and of their contribution to business processes and services, and hence of the broader effects of a localized IT investment decision.

6   Business Requirements Modeling

Knowing what the overall IT strategy and resulting operating model is, is only a first step. Next, we have to make these strategic choices more concrete and define the resulting business goals and requirements.

A desired organizational and/or technical change requires the investigation of the stakeholders that are involved and their concerns regarding the change. New goals and requirements are identified, or existing ones are changed, to address these concerns. Analysis of these goals and requirements is needed to guarantee consistency and completeness, and to propose one or more alternative architecture designs that realize the goals and requirements. Validation of these alternative designs aims at assessing their suitability and selecting the best alternative.

This way, business requirements capture the motivation and rationale behind (the design of) enterprise architectures. Furthermore, architecture artifacts, such as business services, processes and supporting software applications, are related to the (high-level) goals and requirements they originate from. Or put in another way, goals and requirements can be traced towards the architecture artifacts that realize them. This traceability between goals and requirements on one side and architecture artifacts on the other side is important to valuate these artifacts. In the context of portfolio management, the valuation of artifacts that represent or require IT support is of particular interest. The valuation of some artifact in terms of the allocation of costs and benefits may largely depend on the goals and requirements to which the artifact contributes.

Problem chains link requirements engineering to enterprise architecture. This is illustrated in Figure 4. The *why* column represents the problem-oriented view and defines the business needs, goals, requirements and use-cases that should be addressed.

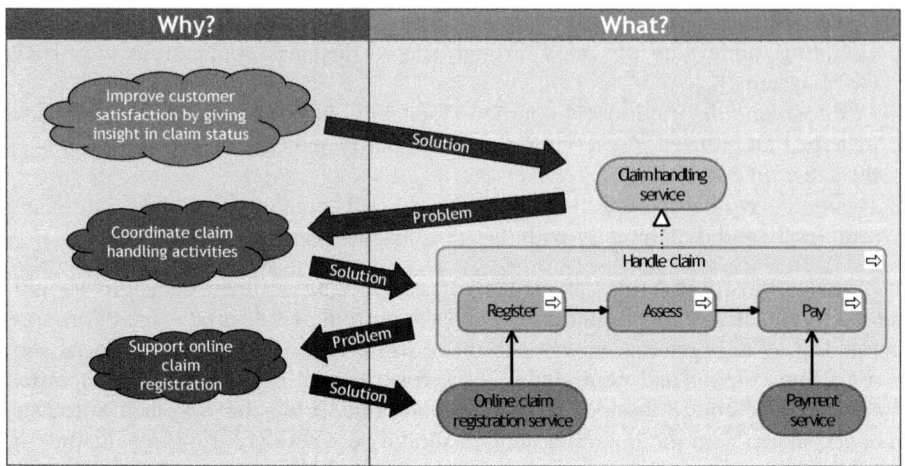

Fig. 4. Relation between requirements engineering and enterprise architecture

The *what* column represents the solution-oriented view in terms of enterprise architecture artifacts, such as services, processes and applications. These architecture artifacts define *what* the enterprise must do to address the business needs, goals, requirements and use-cases. At the same time, these requirements engineering artifacts motivate and justify *why* the enterprise architecture is defined the way it is.

Quartel et al. [14] present a method and modeling language (ARMOR), as an adjunct to the ArchiMate language [9, 10], which support business requirements engineering as outlined above. The intention is to incorporate such concepts in a future version of The Open Group's ArchiMate and TOGAF standards [15].

In general, requirements engineering starts with some organizational goal that needs to be addressed. This issue cannot be approached 'from scratch', but has to take the current organization into account. This means that any requirement or goal should be defined 'relative' to the as-is architecture in order to address the change. In this situation, the as-is architecture acts as a frame of reference for (problem-oriented) requirements engineering. Subsequently, a to-be architecture is designed that realizes a solution for the requirements and goals. In this situation, the to-be architecture is considered a design artifact that results from (solution-oriented) requirements engineering.

Requirements engineering can be further decomposed into three steps:

- **Problem investigation**, which focuses on the problem, i.e., the organizational change, by identifying and analyzing its cause in terms of the involved stakeholders and their concerns, and by eliciting goals to deal with the change. Goals are structured and analyzed for consistency and completeness. In addition, the impact that goals may have on each other is analyzed, e.g., to detect conflicts.
- **Investigation of solution alternatives**, which refines the goals in order to find possible solutions to realize them. The (impact) analysis from the previous step typically triggers the identification and elaboration of alternative solutions. The possible solutions are guided and constrained by architecture principles, which are specialized into requirements for the specific problem and solution at hand.

– **Solution validation**, which validates alternative solutions and chooses the 'best' among them. This choice is amongst others influenced by the impact each solution has on the desired goals, i.e., how well the solution satisfies the goals, and by how well each solution conforms to the architecture principles that apply.

These steps constitute a generic requirements engineering cycle that can be repeated at successive phases in the development of some enterprise architecture. Furthermore, the identification, analysis and refinement of solution alternatives in the second step may be repeated as well, leading to 'sub-cycles'. In this paper, we will not go deeper into this topic, but merely use these ideas in an example below. The interested reader is referred to the aforementioned publications.

7 Calculating Portfolio Value

In the previous sections, we addressed the business strategy and requirements that provide the context for determining the value of an organization's IT portfolio. In our method, the value of the IT portfolio is related to the way in which IT projects and applications support these strategic goals and requirements. To assess such a portfolio, the contributions of its various elements to the goals of the organization must be determined. Note that this method does not provide a direct link with the effects of IT projects, and hence of the organization, on the outside world; rather, it assesses the contribution of the IT portfolio to the organization's goals, which in turn may contribute to such external effects.

A contribution can be divided into two elements: its importance to a business goal and the quality or effectiveness in supporting that goal. The value of an organization's IT portfolio thus depends on the contribution that its constituent elements provide to the business.

Fortunately, we need not devise a method for calculating these business contributions and the value of IT portfolios from scratch. Bedell's method [1] does precisely that. It answers three questions:

1. Should the organization invest in information systems?
2. On which business processes should the investment focus?
3. Which information systems should be developed or improved?

Moreover, it provides a neat relation to the enterprise architecture of the organization, as we will show in Section 7.1; thus, it relates IT strategy to both architecture and design, and to investment decisions.

The underlying idea of the method is that a balance is needed between the level of *effectiveness* of the information systems and their level of *strategic importance*. Investments are more crucial if the ratio between the effectiveness of an information system and its importance is worse.

In order to calculate this ratio, the following information needs to be determined:

– The importance of each business process to the organization.
– The importance of information systems to the business processes.
– The effectiveness of the information systems to the business processes.

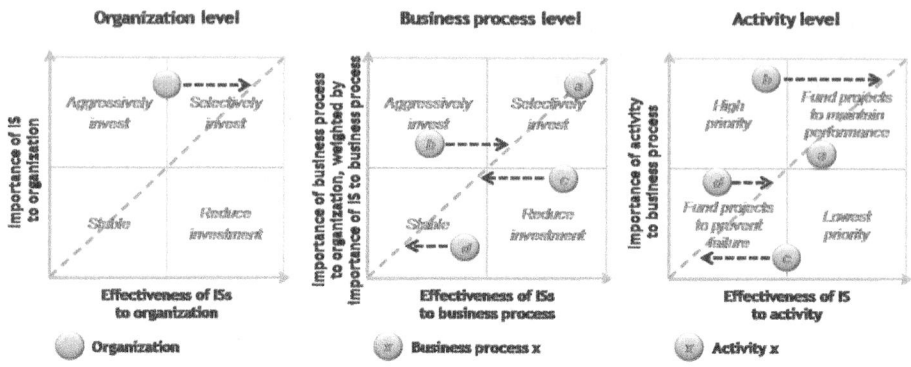

Fig. 5. Investment portfolios

Based upon this information, three portfolios are calculated: for the organization as a whole, for its business processes, and for the information systems that support these processes. Figure 5 depicts an example of all three portfolios and associates a general investment decision to each quadrant of the portfolios. A dashed arrow points to the ideal position of some organization, business process or information system (IS) in the portfolio.

The prioritization of investment proposals is determined by the contribution of each information system, which is defined as the product of its importance and the projected improvement of its effectiveness. In addition, the value of the investment can be evaluated by calculating a so-called project-return index. This index relates the contribution of the information system to the development costs.

7.1 Foundation

Bedell's method is well-suited to be used in combination with enterprise architecture models. Figure 6 depicts the architecture elements on which the method operates: a business actor that represents the organization as a whole, the business processes of the organization, the activities that are performed by the business processes, and the information systems that support these activities. The architecture elements are represented in the ArchiMate language [9].

For convenience, the 'used by' relation is used to relate the architecture elements, except for the aggregation relation between an individual 'Information system' (represented as an application service) and the collection of (all) 'Information systems'. Bedell assumes the following restrictions on the architecture model: (i) a business process may comprise multiple business activities, but a business activity contributes to only a single business process, and (ii) a business activity is supported by a single information system (represented as an application service), and an information system supports only a single business activity. We have generalized the method to overcome these restrictions.

The names that are annotated to the 'used-by' relations in Figure 6 represent the variables that need to be determined as input to the calculation of the investment portfolios as depicted in Figure 5:

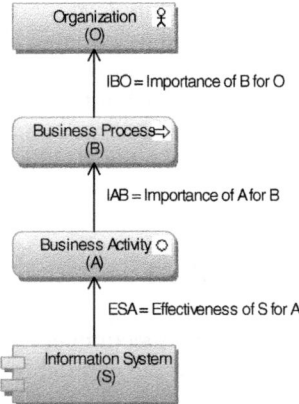

Fig. 6. Bedell's method and enterprise architecture

- IBO = the current **I**mportance of some **B**usiness process to the **O**rganization;
- IAB = the current **I**mportance of some **A**ctivity to some **B**usiness process;
- IIB = the potential **I**mportance of **I**nformation systems to some **B**usiness process;
- ESA = the current **E**ffectiveness of some Information **S**ystem to some **A**ctivity.

7.2 Plotting Portfolios

The information obtained from computing these indicators can be shown graphically, as illustrated by the figure below. This type of plot is familiar to anyone who knows the business value – technical value diagrams used by, for example, the ASL methodology, in particular its Application Lifecycle Duration Measurement Method (ALMM) [15].

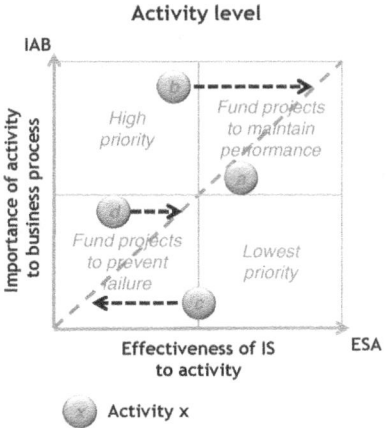

Fig. 7. Example of an activity level portfolio

Figure 7 depicts an example of an activity-level portfolio. The importance of an activity to a business process is represented by variable IAB at the y-axis. The effectiveness of a single information system in supporting an activity is represented by variable ESA at the x-axis. Similar plots can be made at the business process and organizations levels. In the next section, we will show how such portfolios can be computed from a model of the enterprise architecture in combination with input on the quality and importance of information systems.

8   Implementation of the Method

To demonstrate the use of enterprise architecture and business requirements modeling for portfolio valuation, we implemented the method of Bedell in the BiZZdesign Architect tool[1]. This tool enables the modeling and analysis of enterprise architectures in ArchiMate. In addition, this tool supports ARMOR, the aforementioned extension of ArchiMate for modeling business goals and requirements.

The implementation of Bedell's method is illustrated by means of an example, the fictitious enterprise PROFIT. After presenting the enterprise architecture of PROFIT, we valuate its IT and project portfolio following the method given in the previous sections.

8.1   Example: PROFIT Insurances

PROFIT is an average sized financial service provider, specialized in insurance packages, such as life insurances, pensions, investments, travel insurances, damage insurances and mortgages. We limit the scope of this example to a single, generic product: insurances.

Figure 8 depicts a product view on PROFIT's enterprise architecture showing the Insurance product, the business services that are provided by this product, and the business processes that realize these services. In addition, the importance of the business services for the product, denoted by variable ISP, is represented by a value in the range 0..10 next to the aggregation relation that links the business service to the product.

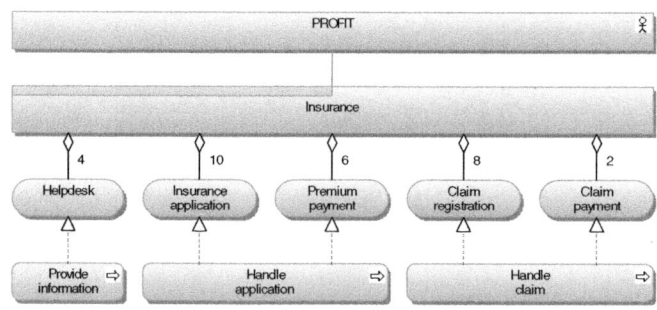

Fig. 8. Product 'Insurance' of PROFIT

---

[1] http://www.bizzdesign.com/index.php/tools/architect

Variable IBO can be derived from ISP as follows: IBO = Max$_S$/Avg$_S$/Min$_S$(ISP), where S ranges over the services that are supported by B. In this case O represents PROFIT and we do not valuate the importance of the Insurance product (P) to PROFIT. This example shows that the application of Bedell's method is not limited to the architectural levels of Figure 6, but may include additional or alternative levels.

Figure 9 depicts the decomposition of process 'Handle claim' into business activities and applications that support these activities. In addition, values for variables IAB and ESA are depicted next to the aggregation relations and used-by relations, respectively. Similar decompositions have been made for processes 'Handle application' and 'Provide information'.

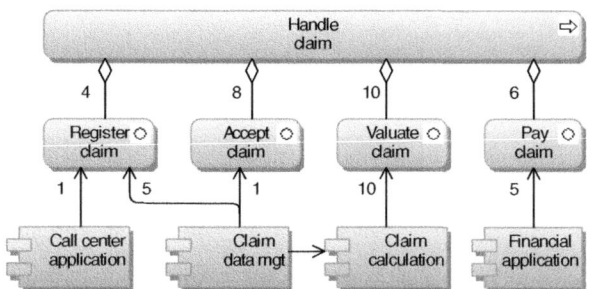

Fig. 9. Business process 'Handle claim'

8.2 Valuation of the As-Is IT Portfolio

Based on the as-is enterprise architecture and the provided values for ISP, IAB and ESA, a portfolio can be calculated that shows the importance of IT support against its effectiveness. Such a portfolio may be calculated for different architectural levels. Bedell distinguishes three levels: the organization as a whole, its business processes, and the activities within a business process.

Figure 10 depicts the portfolio for the activities of business process 'Handle claim', as generated by the BiZZdesign Architect tool. This portfolio shows that the importance and effectiveness of IT support for activity 'Valuate claim' is in perfect balance. For activities 'Pay claim' and 'Register claim' the effectiveness of IT support may need improvement to achieve an optimal balance. However, investments are definitely needed for activity 'Accept claim', since it is considered important but the effectiveness of IT support is (very) low.

As another example, Figure 11 depicts the portfolio of PROFIT at business process level. The effectiveness of IT support for processes 'Handle application' and 'Handle claim' needs some improvement. Instead, the effectiveness of IT support for process 'Provide information' is higher than 'required' given its importance. This means that PROFIT could consider to de-invest in this process, e.g., by cutting budget for development and/or maintenance.

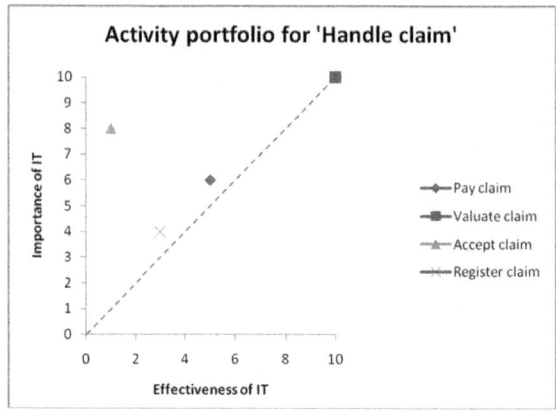

Fig. 10. Activity portfolio of business process 'Handle claim'

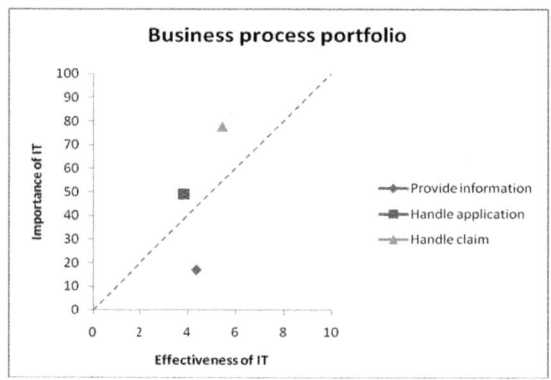

Fig. 11. Business process portfolio

8.3 Project Development

Based on the portfolios that are calculated for the as-is architecture, necessary and optional improvements for IT support can be identified. Subsequently, projects are proposed to implement these improvements. For example, Figure 12 illustrates a project to increase the effectiveness of the 'Claim data mgt' application. It shows the part of the architecture affected by the project. Furthermore, it shows the to-be situation, including the added effectiveness compared to the as-is situation, represented by the values '+5' and '+4' for variable ESA. In this case the effectiveness of IT support for activities 'Handle claim' (which scored low in Figure 10) and 'Register claim' is improved. The improvement for the latter process may be unnecessary, but comes more or less for free since it is supported by the same application as 'Accept claim'.

PROFIT proposes five additional projects that vary from improving the effectiveness of current applications, replacing applications by better ones, to adding new applications. Figure 13 depicts a project that introduces a new application 'On-line assessment' to support a new activity 'Second opinion' in process 'Handle claim'.

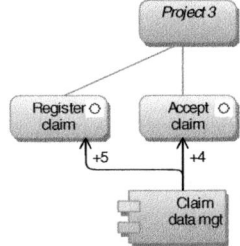

Fig. 12. Project to improve IT support for 'Accept claim'

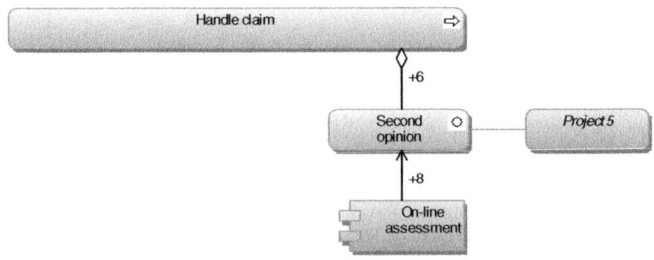

Fig. 13. Project to introduce a new application

8.4   Valuation of Projects and the To-Be IT Porfolio

In general, not all proposed projects can be selected. Therefore, projects are ranked based on their so-called project return index (PRI). This index represents the ratio between the effectiveness of IT support that is added by a project and its costs, with

$$PRI = Added / Costs$$
$$Added = (EIA_{to-be} - EIA_{as-is}) * IAO$$

Figure 14 plots the ratio for each of the proposed projects.

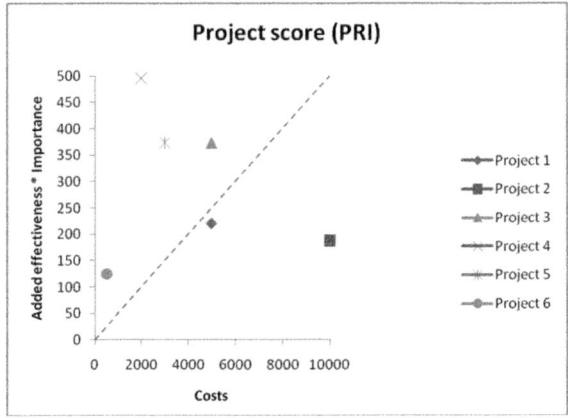

Fig. 14. Project return index of proposed projects

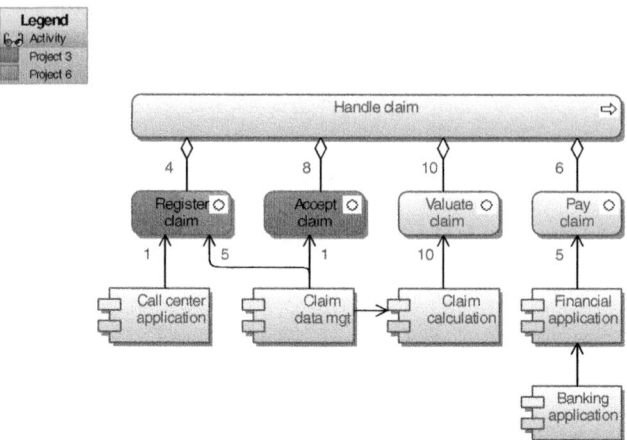

Fig. 15. Project overlap view

Following a naïve approach, projects with the highest ranking could be picked until the available budget is allocated. However, some projects may overlap because they affect (partly) the same applications, activities or processes. A specific viewpoint has been implemented in BiZZdesign Architect to detect such overlaps. For example, Figure 15 depicts the projects, and their overlap, affecting process 'Handle claim'.

The detection of overlaps between projects may (should) motivate the redefinition of these projects to obtain a better separation of concerns and alignment. However, it is not likely that project overlaps can be avoided completely. Therefore, we also support the calculation of the added value (effectiveness) that is provided by a selection of projects.

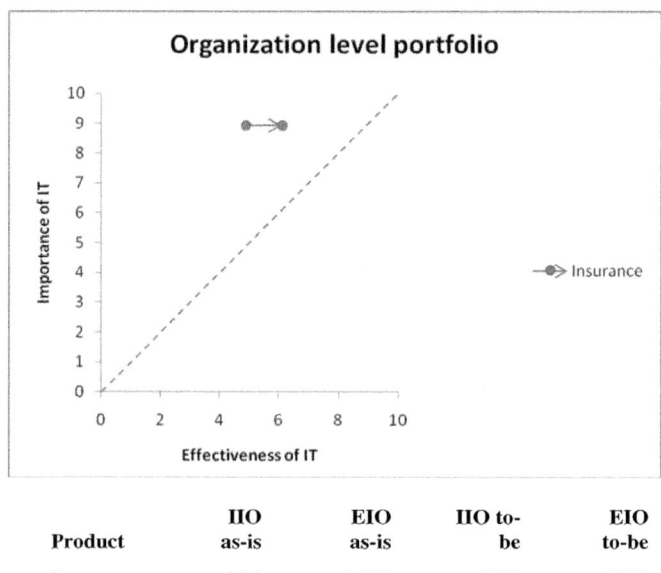

Product	IIO as-is	EIO as-is	IIO to-be	EIO to-be
Insurance	8.94	4.894	8.94	6.103

Fig. 16. Added value for a selection of projects

For example, Figure 16 depicts the added effectiveness at organization level for a selection of three projects (numbers 3, 4 and 5). This can be repeated for different selections of projects. In this way, the added effectiveness of different sets of projects can be compared to find the set that offers the best investment in terms of added value per unit of money.

9 Expressing and Assessing Value

In Bedell's method, as explained in the previous sections, an information system is considered effective when it is cost-effective, has high technical quality and is functionally appropriate. It is considered strategically important when the activities it supports are crucial to a business process or the organization in obtaining its strategic objectives. However, the assessment of these properties is rather subjective and lacks concrete guidance. Hence, we need more concrete measurements of the properties of value of the IT landscape.

To measure the effectiveness of IT in supporting a business goal, we need measurable performance indicators. This implies that an organization should define for each business goal the key performance indicators (KPIs). These KPIs can be obtained by decomposing a goal into sub-goals, possibly iteratively, until measurable sub-goals are obtained. For example, the business goal 'Improve helpdesk' may be decomposed into the sub-goals 'Reduce waiting to a maximum of 1 minute' and 'Improve quality of answers'. In this case, the first sub-goal is measurable, whereas the second one may need further decomposition to become measurable.

As explained before, the ARMOR language [14] can be used to model the decomposition of business goals into KPIs. For example, the left side of Figure 17 models 'Customer satisfaction' as a concern of stakeholder 'Board' and having the assessment 'Customers complain about helpdesk', which is decomposed into the sub-assessments (sub-complaints): 'Long waiting queues' and 'Inadequate answers'. Business goal 'Improve helpdesk' addresses the assessment 'Customers complain about helpdesk', such that each sub-goal addresses a sub-assessment.

For assessing a portfolio, specific measures are needed as input for decision making. These measures or KPIs should be derived from the business goals. Bedell's method uses 'importance' and 'effectiveness' as major criteria, which are both single measures related to an application or business process. The notion of 'effectiveness' (or rather 'quality') is broad and it depends on the value center approach that the organization chooses. For a service center approach, for example, customer satisfaction is an important criterion; the effectiveness with which a service supports this may depend more on aspects such as usability. For a cost center, low maintenance and efficient usage of resources is important, and for an innovation center, flexibility of a system is essential to obtain an effective support of future capabilities.

Although the scope of the concept of effectiveness is large, the various views can all be related to concepts of IT quality that are addressed in the ISO 9126 standard for software quality [17]. Although this standard was originally intended for classifying various types of requirements posed to an information system before it is built, the attributes can also be used to assess its qualities after it has been constructed.

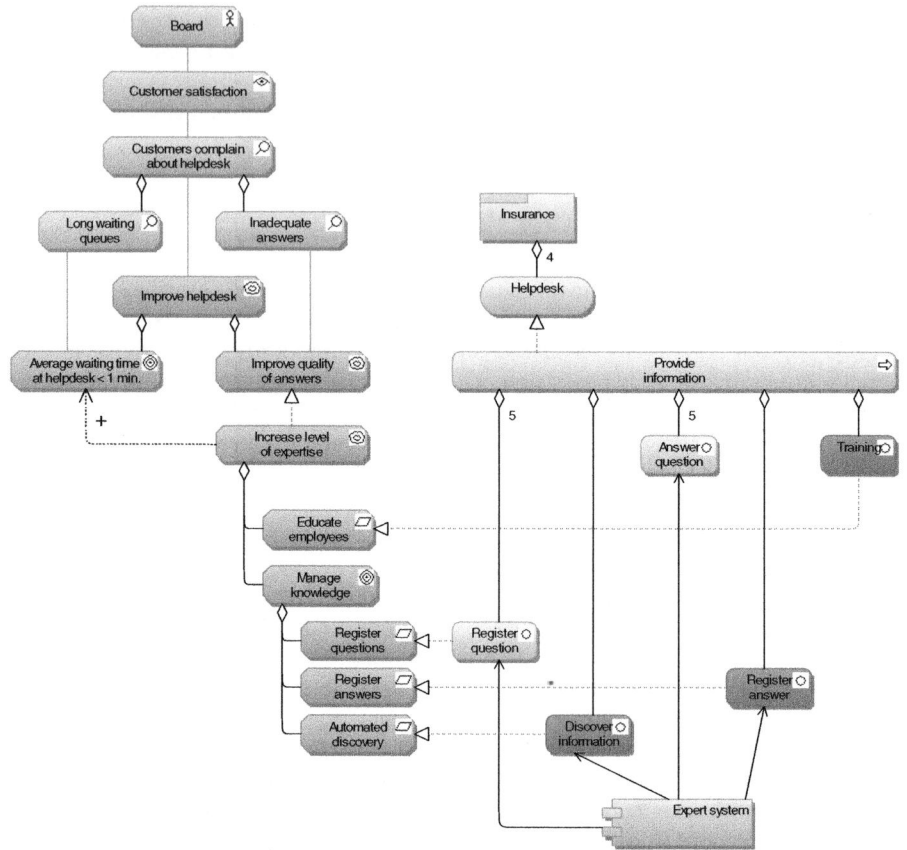

Fig. 17. Relations between stakeholders, goals and architecture artifacts

The notion of 'importance' is more difficult to address. Although methods such as ASL [15] provide questionnaires to investigate the business value of applications (see below), much of this value is dependent on, for example, the value of the information a system or service provides to the business, the value of future opportunities opened up by IT, or the value of customer satisfaction created by a user-friendly system.

An important category of indicators related to importance addresses risk. In the value center approach, 'risk propensity' is an important factor in the type of value center. The cost and service centers aim for low-risk operations, whereas the profit and investment centers allow for higher risks in order to obtain possible (but uncertain) gains from future business opportunities. There are also risks concerning failure of projects.

9.1 Measuring Importance and Effectiveness

For assessing the IT portfolio, specific measures are needed as input for decision making. Bedell's method uses 'importance' and 'effectiveness' as major criteria, which are both single measures related to an application or business process. Importance and effectiveness are measured by means of a 10-point scale, with a (verbal) meaning attached to each level. Other methods use measures that are composed of multiple aspects. One of those methods is the Application Services Library (ASL) methodology [15], in particular its Application Lifecycle Duration Measurement Method (ALMM). The ALMM measures the life cycle duration of applications by determining their current Business Value (BV) and the Technical Value (TV) and then by estimating the development of the BV and TV in the future, assuming a continuation of the current IT policy. Business Value and Technical Value are both composite measures. The Business Value is defined as "the extent to which an application supports the operating processes." This appears to be very close to Bedell's definition of "strategic importance": a system is strategically important when the activities supported are crucial to the organization or business process in obtaining its strategic objectives. The Technical Value is defined as "the extent to which an application can be efficiently adapted to changing circumstances." This is close to Bedell's notion of "effectiveness": a system is regarded to be effective when it is cost-effective, has a high technical quality, and is functionally appropriate. Especially the latter can only be assured if the system can be adapted to changing functional demands.

To assess the BV and TV of an application, ASL uses an extensive questionnaire to assess 22 parameters for the application, 11 parameters for the business value and 11 parameters for the technical value. Each application is assessed by questioning management, users, functional administrators, and technical administrators on aspects such as degree of coverage of the required functionality, importance to other systems, continuity of the supplier, accessibility of user documentation, quality of the data, logical coherence between the functionality of the application and the operating processes, how much corrective maintenance is needed, and availability of the necessary technical knowledge. For each parameter, a score of good (2 points), moderate (1) or bad (0) is given. An overall score is calculated as a weighted average of the BV and TV related aspects, where weights are defined beforehand and depend on the organization's specific characteristics.

Both approaches (Bedell's with a single measure and ASL with a composite measure) have advantages and disadvantages. The accuracy of composite measures may be higher because the aspects have a smaller scope and hence are easier to asses. However aspects have to be relevant and should be properly weighted. Composite measures take more measurement time, especially when more people are involved. Which approach is relevant depends on the scope of the organization and the skills, experience and expertise of the people that are involved in the valuation. For example when rough indications are sufficient, approaches such as used by Bedell with single measures might be appropriate.

The approaches described above focused on human ratings that are averaged or aggregated to obtain a specific measure. Next to that there are other ways of assessing specific aspects of IT or business processes, that can be used as alternatives.

With respect to effectiveness or technical value, more concrete measures are available. Either performance measures obtained by monitoring actual IT usage or measures that are based on an analysis of the IT architecture can be used. The quantitative analysis techniques developed in the ArchiMate project can serve as a basis for the latter [18]. For example, *utilization* is the percentage of the operational time that a resource is busy. Such a measure can be obtained by monitoring and used as a measure of the effectiveness. When the architecture model is augmented with attributes such as the number of service calls from each activity, arrival frequencies for the various business processes, et cetera, it is possible to compute a composite estimate of the utilization of a system.

As an example of using the architecture to derive qualitative indicators, consider the ASL ALMM parameter for Strategic Context (SS). Strategic Context is the measure in which an application provides direct contribution to the policy goals of the organization. For the assessment of this indicator, the nature of the processes that are supported by the application is considered: the greater the importance of the process for the organization, the more important the application. Such a measure could be calculated on the basis of the ArchiMate EA model, as follows:

$$SS(app{:}ApplicationComponent) =$$
$$\text{MIN}(2, \text{MAX}(\{ \text{IMPORTANCE}(p) \mid p{:}BusinessProcess\ app.realizes.used\_by \}),$$
$$\text{where IMPORTANCE}(p{:}BusinessProcess) =$$
$$\text{SIZE}(\{goal{:}Goal \mid <p,goal> \in contributes\_to$$
$$\text{AND } <p,goal>.value > 0 \text{ AND } goal.type = \text{'}strategic\text{'}\}) .$$

For example, this formula can be used to calculate the strategic context of application component 'Call center application' in Figure 9. First, the business processes are determined that use the application. The formula assumes that each application component is used by a process via some application service it realizes; as represented by condition '*app.realizes.used_by*'. However, Figure 9 omits the level of application services, but considers the level of business activities instead. The formula could (and should) be generalized to accommodate for these situations. In this case, application of the formula returns the process 'Handle claim' (as can be derived from Figure 9), and the processes 'Handle application' and 'Provide information' (these models are not shown in the paper for brevity). Subsequently, it is determined how many of these processes contribute to a strategic goal. The contribution of some process to a goal can be represented using the contribution relation of the aforementioned ARMOR language. In this case, the value of the contribution should be larger than zero. Finally, the value of the strategic context is defined as being equal to the number of 'strategic processes' with a maximum value of 2 (since an ALMM parameter can only have the value 0, 1 or 2).

Measuring importance or business value requires insight in the ways in which an IT system contributes to the business. This value can lie in many different aspects, encompassing, for example, the timely and accurate information that the system delivers as input for business decisions, the customer satisfaction and return business created through its user-friendly interface, or the value of future opportunities opened up by IT. This is still largely uncharted territory.

9.2 Risk Analysis

Another highly important category of indicators addresses risk. Risk in general is one of the criteria that managers base their decisions on. 'Risk' is often defined as the effect of uncertainty on business goals (e.g. in the ISO 31000 guide [19]). In the value center approach, 'risk propensity' is an important factor in the type of value center.

The cost and service centers aim for low-risk operations, whereas the profit and investment centers allow for higher risks in order to obtain possible (but uncertain) gains from future business opportunities. There are also risks concerned with failure of projects. So how should we measure 'risk'?

The ISO 31000 family of standards addresses risk management procedures and risk assessment techniques. The first member of this family, ISO 31000:2009, was published as a standard on the 13th of November 2009, and provides a standard on the implementation of risk management [19]. ISO/IEC 31010:2009 [20] provides guidance on selection and application of systematic techniques for risk assessment. The ISO/IEC 31010 standard describes about 30 different techniques for risk assessment, comprising such diverse approaches as fault mode and effect analysis, hazard and operability studies, scenario analyses, and Delphi studies. However, the ISO standards are highly generic and can be applied to risk management in many different fields; in our case of IT related risks, only a subset of these methods will be applicable. This is something we intend to investigate in the future.

9.3 Other Measures

Next to effectiveness, importance and risk other criteria are relevant for the valuation of IT systems or IT projects. Among those not yet discussed are customer satisfaction, information value, and flexibility and scope of future opportunities.

Customer satisfaction is relevant when companies maintain a service center approach or when there are specific requirements for service levels. The contribution of IT to ultimate customer satisfaction is typical composite measure, involving aspects such as:

- support to the speed of activity realization;
- service levels: quality, up-time;
- usability: ease of use with respect to activity;
- support of customer processes;
- provided accuracy of information (e.g. content);
- provided accuracy of information delivery (e.g. process, like right time, right place, right person/system).

Much of the value of an IT system is determined by the value that the business attaches to the information the system delivers. Capgemini's Global CIO Report [21], for example, estimates that 80% of IT value is information value, and only 20% deployment value. Timely and accurate information on which important business decisions can be based is a vital asset for any company. However, no concrete measurements are currently available to assess this information value in a readily applicable way. Some example aspects that could be included in an indicator of information value are:

- value of business decisions based on the information, for example in terms of financial risk of wrong decisions;
- potential for cross- and upselling by using customer data effectively, or conversely, business opportunities lost due to missing information;

- competitive advantage of having the information available within a certain time frame, for example earlier than competitors;
- compliance with applicable laws and regulations;
- cost of assembling the information by hand.

Assessing future opportunities requires even more complex measures. On the one hand, scenario analyses may serve to gain insight in the various kinds of opportunities that may arise; on the other hand, an analysis of the architecture may reveal what the effort is that seizing this opportunity will take. Next to measures of flexibility and changeability as described before, many other aspects should be taken into account as well. Given the diverse nature of possible business opportunities, we do not go into detail here.

10 Valuation Profiles

A valuation profile operationalizes the value center concept from Venkatraman and tailors it to the organization at hand. The definition of a valuation profile involves the following steps:

1. Define the value centers;
2. Define for each center the relevant concerns and business goals;
3. Define the KPIs for each business goal;
4. Define the importance of business goals.

Each of these steps requires choices to be made by business and IT management. The choices concern the strategy of the organization, the business values that are considered important, the goals that are set to implement the strategy and realize these values, the relative importance of these goals, and the way they are measured. These choices will be subject to debate that may involve various issues, including political ones. The purpose of this work is not to rationalize this process, but to make it transparent and to facilitate reasoning and analysis.

In addition, separate budgets may be allocated to each value center to distribute investments over the selected sources of business value. For example, management could decide to allocate the following percentages of the total IT budget to the four value centers, respectively: 40%, 35%, 15% and 10%. This means that management is willing to invest in each value center, but emphasizes IT (projects) that support the business goals of the cost and service centers.

Each value center can be characterized by a number of concerns and business goals. For example, the service center may be concerned with: customer satisfaction and internal service levels. The (periodic) assessment of these concerns lead to the definition of business goals that address these assessments. These goals are used to valuate IT artifacts and projects within a certain value center, by assessing the contribution of IT to these goals.

For example, assessment of the concern of customer satisfaction may reveal that customers complain about the helpdesk, as modeled in Figure 17. This leads to the definition of the business goal 'improve helpdesk' and its decomposition into more concrete sub-goals and requirements. IT projects that contribute to these goals and requirements, e.g., by implementing parts of the 'Expert system' that supports the business activities 'Discover information' and 'Register answer', are likely to receive funding from the service center budget. The amount of funding will depend amongst others on the number of 'competing' projects that also contribute to (improved) customer satisfaction, and the value of their contribution.

Similarly to Bedell's method, we split the contribution of IT to business goals into:

1. the *effectiveness* of IT support, and
2. the *importance* of IT support.

10.1 Valuation Based on Business Goals

The method of Bedell does not consider the calculation of portfolios for different value centers. Instead, it assumes general criteria for assessing the importance and effectiveness of IT support. The valuation profile as described above characterizes a value center by a number of, possibly prioritized, business goals. The IT portfolio assessments of Bedell can be calculated for a specific value center by valuating IT artifacts against the business goals that are associated with this center.

Using such techniques, portfolios can be calculated for each value center in the valuation profile to assess the importance and effectiveness of IT support at different architecture levels. This may lead to different rankings of IT projects for each value center. This means that other projects may be selected when compared to the case in which no value centers are distinguished. Which projects are candidates to be selected, depends on the budgets that are allocated to the value centers (or the relative priorities of their business goals). However, management can still decide to divert from the ranking and allocated budgets. An important reason for this is that social and political influences are not reflected in the approach. In this respect, we remind the reader that the purpose of our approach is to make the valuation criteria and techniques within an organization clear, precise and transparent, and to enable their systematic application. The purpose is not to enforce decision making.

10.2 Linking Indicators to Value Center Approaches

The relative importance of various indicators in assessing an IT system (landscape) depends on the chosen valuation profile. To provide the connection between the IT strategy and value center approach within the context of portfolio management, we have investigated a first mapping between the four value centers and the specific indicators that are most relevant for these centers. Table 1 provides a first indication of especially important aspects (apart from functionality, which is of course important in all cases) for the four types of value centers.

For example, if we want to assess the effectiveness of an IT service within the context of an investment center, we should put a stronger weight on its interoperability, maintainability and portability, since the goal of an investment center is to build future capabilities and the service should support the necessary changes. In the context of a cost center, focused on operational excellence, efficient and reliable operation, and low risk are more important.

Table 1. Value centers and indicators

Value center type	Aspects
Service center	*Usability*: if customers have direct interaction with IT services, usability is of great importance to customer satisfaction. *Market risk*: no risky new business ventures, but optimally serving clients with existing products and services. *Operational risk*: should be kept low, to ensure uninterrupted service to customers. *Reliability*, in particular availability, and performance: the customer experience depends on a service to operate as and when advertised, with adequate performance and without unfortunate surprises. *Information value*: the value of the information provided by the services is an important part of the business capability that a service center supports.
Cost center	*Efficiency*, in particular resource behavior: operational excellence requires low operational costs and efficient use of resources. *Risk*: operational, market and project risks should be kept low, to avoid costs associated with outages, runaway projects or failed new business ventures. *Reliability*: to minimize risk, the IT landscape should be very reliable and require minimal maintenance. *Information value*: the value of the information provided by the service to the business puts an upper bound on its costs.
Investment center	*Maintainability*, in particular changeability: to build future capabilities, IT systems need to be flexible and amenable to experimenting with new functionality. *Portability*, in particular adaptability: a changing or uncertain future environment may require the service to function under different (possibly unknown) circumstances. *Functionality*, in particular interoperability: in future, new environments, the system might need to be linked to other systems to provide new, joint capabilities. *Future business opportunities* opened up by the service.
Profit center	*Maintainability*, in particular changeability and reusability: to compete with best-in-class vendors, IT systems must be easily adaptable to stay ahead of the competition, and changes should require minimal cost and effort. *Efficiency*, in particular resource behavior: achieving maximum profit and operational excellence requires low operational costs and efficient use of resources. *Usability*, in particular attractiveness: a profit center focuses on delivering IT services in an external marketplace. The customers of these services should be offered an attractive experience compared to the competition.

11 Related Work

While reviewing the literature on IT valuation approaches, we found several interesting relations to EA. Architectures are seen as critical to managing the IT assets in complex organizations [22]. EA provides insight into what projects are required by doing a gap analysis between the as-is and to-be situation. This occurs in response to business drivers and strategic intentions.

TOGAF is a framework for EA that is defined by the Open Group [8]. A pivotal part of TOGAF is a methodology for developing the architecture design, which is called the Architecture Development Method (ADM). TOGAF ADM's phase F (Migration Planning) allows architecture changes to be aligned and concerted with cost benefits analysis.

The Architecture Tradeoff Analysis Method (ATAM) [23] provides software architects with a framework for understanding the technical tradeoffs they face as they make design or maintenance decisions. But the biggest tradeoffs in large complex systems usually have to do with economics, and the ATAM does not provide any guidance for understanding these economic tradeoffs.

The Cost Benefit Analysis Method (CBAM) [24] helps software architects to consider the return on investment of any architectural decision and provides guidance on the economic tradeoffs involved. The CBAM guides the stakeholders to determine the costs and benefits associated with the architectural decisions that result in the system's qualities, mainly performance, availability, security, and modifiability. Given this information, the stakeholders can then reflect upon and choose among the potential architectural decisions.

Val IT [4] is a relatively new standard that provides a framework for the governance of IT investments, which mainly focuses on the principles and processes that help business managers to get business value out of their IT investments. It is closely related to the COBIT IT governance framework [25].

The idea of combining architecture with financial or economic valuation methods is not new. Nord et al. [26] propose to integrate the ATAM with the CBAM as a way to bridge architecture and investment decisions. Wieringa [22] makes a similar case for combining EA with a balanced scorecard or information economics approach to reach new synergies. However, these proposals do not provide much detail on how the actual valuation should be performed.

The closest related work is that by Johnson et al. [27], who have developed a method for creating enterprise architecture meta-models suitable for a specific kind of architecture-based analysis and the subsequent performance of the analysis using probabilistic relational models. An example of the method where ArchiMate is extended to analyze the impact of IT on business goals, such as flexibility, efficiency and effectiveness, is presented in [28]. Instead of the ARMOR extension, extended influence diagrams are used to relate EA concepts to organizational goals. Another difference is that we currently do not support reasoning with uncertainty, but this could be added in future.

As our analysis of the existing literature shows, very little research is currently addressing the full relationship between business benefits, IT strategy, enterprise architecture and portfolio selection. With our work on architecture-based valuation we aim to fill this gap.

12   Conclusions and Future Work

In this paper we have presented an approach for IT portfolio valuation that uses enterprise architecture extended with business requirements modeling as a basis. The approach borrows ideas from Bedell, i.e., the decomposition of the value of IT into the importance and the effectiveness it provides to the business, and from Venkatraman, i.e., the use of a valuation profile to distinguish different sources of value (value centers) and associated business goals.

The theoretical and technical feasibility of the approach has been demonstrated by means of a prototype implementation in the tool BiZZdesign Architect. The practical applicability of the approach is currently being investigated, which involves case studies and benchmarking within large IT organizations in the public and financial sector.

In addition, the presented approach should be embedded in an overall vision on portfolio valuation and management that comprises all phases of the IT life cycle. Decision making and evaluation of alternatives based on the valuation of an IT portfolio requires an assessment of multiple aspects. An obvious case is the combination of financial aspects (e.g. direct cost, TCO, ROI, NPV) in relation to measures of business and technical value or effectiveness and importance, as described in the previous sections. Established financial instruments such as TCO or ROI calculations do not use the architectural structure and dependencies but do their computations only on the individual elements present in the portfolio. The outcomes of these techniques should of course be taken into account in making IT investment decisions.

Each of these techniques results in some assessment or valuation. These results alone are of course not enough. Given an assessment of the cost, returns and qualities of different alternatives, for example renovating an application, replacing it completely, or leaving it as-is, how can the organization decide upon such a multitude of inputs?

Rather than use a separate method for each of these assessments and combining the results by hand, our ultimate goal is to develop a flexible plug-in architecture for architecture-based valuation methods, in which different criteria can be combined using a central framework for multi-criteria analysis. Our aim is to provide an integral approach that can be implemented in tools for architectural design and analysis, to provide optimal support for architects and IT managers.

Moreover, using these techniques as part of the architectural design process, the value of using enterprise architecture as a foundation for decision making is strengthened. Different design alternatives can be assessed on their contribution to business value and well-informed decisions can be made that take the enterprise-wide effects of changes into account.

To assist IT managers in implementing such an advanced portfolio management approach, a staged approach should be taken, in which the organizational maturity is gradually improved and the instruments used fit with the current maturity level. Embedding portfolio management techniques and instruments within an integrated portfolio management and EA maturity model (akin to the CMMI [29]) may assist responsible managers to guide their organization's development.

Acknowledgment

This paper results from the ArchiVal project, a collaboration between Novay, the Dutch Tax and Customs Administration, and BiZZdesign, aimed at developing architecture-based methods for IT portfolio management. For more information: http://www.novay.nl/okb/projecten/archival/7240.

References

1. Schuurman, P., Berghout, E.W., Powell, P.: Calculating the Importance of Information Systems: The Method of Bedell Revisited, CITER WP/010/PSEBPP, University of Groningen, Sprouts Working Papers on Information Systems (June 2008), http://sprouts.aisnet.org/8-37
2. Jeffery, M., Leliveld, I.: Best practices in IT portfolio management. MIT Sloan Management Review 45 (2004)
3. Archer, N.P., Ghasemzadeh, F.: An integrated framework for project portfolio selection. In: Dye, L.D., Pennypacker, J.S. (eds.) Project Portfolio Management Project Portfolio Management, Center for Business Practices, West Chester, pp. 117–133 (1999)
4. ITGI: Enterprise Value: Governance of IT investments: the Val IT Framework 2.0. IT Governance Institute, Rolling Meadows, Illinois (2008), http://www.itgi.org/valit/
5. Venkatraman, N.: Beyond Outsourcing: Managing IT Resources as a Value Center. MIT Sloan Management Review 38(3), 51–64 (1997)
6. Venkatraman, N.: IT Agenda 2000: Not fixing technical bugs but creating business value. European Management Journal 16(5), 573–585 (1998)
7. Ross, J.W., Weill, P., Robertson, D.C.: Enterprise Architecture as Strategy: Creating a Foundation for Business Execution. Harvard Business School Press, Boston (2006)
8. The Open Group: TOGAF^{TM} Version 9, The Open Group, Reading, UK (2009), http://www.opengroup.org/togaf
9. Lankhorst, M.M., et al.: Enterprise Architecture at Work – Modelling, Communication and Analysis, 2nd edn. Springer, Heidelberg (2009)
10. Iacob, M.E., Jonkers, H., Lankhorst, M., Proper, E.: ArchiMate 1.0 Specification. Van Haren Publishing, Zaltbommel (2009)
11. Schekkerman, J.: The Economic Benefits of Enterprise Architecture, Trafford (2005)
12. Slot, R., Dedene, G., Maes, R.: Business Value of Solution Architecture. In: Proper, E., Harmsen, L., Dietz, J.L.G. (eds.) Proc. First NAF Academy Working Conference on Practice-Driven Research on Enterprise Transformation, PRET 2009, held at CAiSE 2009, Amsterdam, The Netherlands, June 11. LNBIP, vol. 28, pp. 84–108 (2009)
13. Ross, J.W., Weill, P., Robertson, D.C.: Enterprise Architecture As Strategy: Creating a Foundation for Business Execution. Harvard Business School Press, Boston (2006)
14. Quartel, D.A.C., Engelsman, W., Jonkers, H., van Sinderen, M.J.: A goal-oriented requirements modeling language for enterprise architecture. In: Proceedings of the 13th IEEE International EDOC Enterprise Computing Conference, Auckland, New-Zealand, pp. 3–13 (September 2009)
15. Engelsman, W., Jonkers, H., Quartel, D.: Supporting requirements management in TOGAF and ArchiMate, White Paper, The Open Group (February 2010)
16. Pols, R., van der Backer, Y.: Application Services Library - A Management Guide. ASL Foundation/Van Haren Publishing (2006)

17. ISO/IEC: Information technology – Software product evaluation – Quality characteristics and guidelines for their use, International Standard ISO/IEC 9126, International Organization for Standardization, International Electrotechnical Commission, Geneva (1991)
18. Iacob, M.-E., Jonkers, H.: Quantitative Analysis of Service-Oriented Architectures. International Journal of Enterprise Information Systems 3(1), 42–60 (2007)
19. ISO: Risk management – Principles and guidelines, International Standard ISO 31000:2009, International Organization for Standardization, Geneva (2009)
20. ISO/IEC: Risk management – Risk assessment techniques, International Standard ISO/IEC 31010:2009, International Organization for Standardization, International Electrotechnical Commission, Geneva (2009)
21. Capgemini: Global CIO Report – Harnessing Information Value: Could you be a digital winner? (December 2009),
    http://www.capgemini.com/insights-and-resources/
    by-publication/global-cio-report/
22. Wierenga, H.: Architectural Information Economics. Via Nova Architectura, August 28 (2009),
    http://www.via-nova-architectura.org/artikelen/tijdschrift/
    architectural-information-economics.html
23. Kazman, R., Barbacci, M., Klein, M., Carriere, S.J., Woods, S.G.: Experience with Performing Architecture Tradeoff Analysis. In: Proceedings of the 21st International Conference on Software Engineering, pp. 54–63 (1999)
24. Kazman, R., Asundi, J., Klein, M.: Making Architecture Design Decisions: An Economic Approach. Technical report. CMU/SEI-2002-TR-035. Carnegie Mellon University, USA (2002)
25. ISACA: CoBiT 4.1. IT Governance Institute, Rolling Meadows, Illinois (2007),
    http://www.isaca.org/cobit.htm
26. Nord, R.L., et al.: Integrating the Architecture Tradeoff Analysis Method (ATAM) with the Cost Benefit Analysis Method (CBAM). Technical report CMU/SEI-2003-TN-038. Carnegie Mellon University, USA (2003)
27. Johnson, P., Lagerström, R., Närman, P., Simonsson, M.: Enterprise Architecture Analysis with Extended Influence Diagrams. Information System Frontiers 9(2-3), 163–180 (2007)
28. Lagerström, R., Franke, U., Johnson, P., Ullberg, J.: A Method for Creating Enterprise Architecture Metamodels – Applied to Systems Modifiability Analysis. International Journal of Computer Science & Applications 6(5), 89–120 (2009)
29. CMMI Product Team: CMMI® for Development, Version 1.2, CMU/SEI-2006-TR-008. Software Engineering Institute, Carnegie Mellon University (August 2006),
    http://www.sei.cmu.edu/reports/06tr008.pdf

Project Portfolio Management in Practice

Michael ter Mors[1], Roel Drost[2], and Frank Harmsen[3]

[1] University of Tilburg, Tilburg
mtermors@gmail.com
[2] Ernst & Young IT Advisory, Utrecht
roel.drost@nl.ey.com
[3] Maastricht University, Maastricht
f.harmsen@maastrichtuniversity.nl

Abstract. This research investigates the approaches that organizations apply to implement project portfolio management (PPM). We have compared theory and practice to find out how organizations can benefit from PPM. The study finds that PPM consists of three tasks: (1) screening, selecting, prioritizing and allocating resources to project proposals, (2) monitoring and reprioritizing running projects, and (3) tracking and managing the realized benefits of projects. We have found a number of opportunities for improvement, since most investigated organizations do not adopt all three tasks. We have found that of the three approaches mentioned in the theory, our respondents use only two. Devoting more attention to the actual outcomes of projects can help organizations to improve their screening and selection process, as well as to take corrective action when intended outcomes are not attained.

Keywords: Enterprise transformation, project portfolio management, business cases, benefits mangement.

1 Introduction

Consider America Online, Inc. (AOL), a U.S. based Internet company. According to Dougherty [1]), AOL grew strongly during the 1990s and its management team in the early 2000s realized their project-based processes were too informal to support further growth. AOL wanted to ensure that its projects reflected strategy and business objectives. They wanted to select high-value projects and find the right mix and balance of projects. AOL also intended to improve accountability through quick and binding decision-making. By 2004, AOL had installed portfolio management teams across its business lines that gave it far more control over its projects and project portfolio. The result was a reduction in the yearly demand of project man-hours from around 200,000 to about 120,000 as well as an increase in return on investment (ROI) of the project portfolio as a whole. So, how did AOL realize a 40% reduction in man-hours while simultaneously improving its portfolio ROI? Essentially, they have asked and answered two questions: what projects should we take on and what projects should we drop?

F. Harmsen et al. (Eds.): PRET 2010, LNBIP 69, pp. 107–126, 2010.

Project portfolio management answers these questions by making an inventory of current and proposed projects and by developing criteria that enable a ranking and comparison of these projects. It is an iterative process that must continually keep track of the project portfolio to ensure fit with business objectives. Taking into account the entire portfolio of projects and interdependencies between projects allows organizations to optimize the contribution of all projects taken together to the overall welfare and success of the organization, as demonstrated by the example case of AOL [2].

Project portfolio management is essential in enterprise transformation, as it enables organizations to manage the transformation in a controlled and justified manner. Typically, enterprise transformations are conducted through a series of projects, programs and activities. Planning and managing these is complicated, and allows for mechanisms that take into account interdependencies, (financial) benefits and control structures.

Research on portfolio management started in finance. Markowitz [3] was among the first to construct a model for securities portfolio selection (dubbed modern portfolio theory). He presented the idea of an 'efficient frontier': an optimal balance of expected returns and variance of returns. Halfway through the 1990s, researchers and practitioners became more interested in portfolio theory geared towards projects [4]. Project portfolio management is defined as "the managerial activities that relate to (1) the initial screening, selection and prioritization of project proposals, (2) the concurrent reprioritization of projects in the portfolio, and (3) the allocation and reallocation of resources to projects according to priority" [5].

Existing literature has pointed out that project portfolio management is important to several business disciplines. McFarlan for instance, argued that companies should create a risk profile of their entire portfolio of Information Technology (IT) projects to maintain a desirable aggregate risk level [6]. According to McFarlan, firms should balance innovative yet riskier projects for future competitive advantage, as well as more conservative projects that support present-day operations. In a New Product Development (NPD) environment, Cooper, Edgett, & Kleinschmidt state that project portfolio management is important as a means to operationalize business strategy (i.e. the products, markets, and technologies that the business wants to focus on) [7]. These decisions direct the business for about five years into the future and products introduced in the past five years generate approximately 32% of companies' current sales. Project portfolio management would help to improve success rates by better aligning projects with the organization's strategy and balancing the portfolio of projects in terms of type and risk. This enables firms to maintain a number of projects in their portfolios that can be resourced effectively, but that is still sufficient to ensure an adequate flow of projects and product introductions [8].

In addition, Archer and Ghasemzadeh [9], Cooper, Edgett, & Kleinschmidt [10] and Blichfeldt and Eskerod [5] argue that project portfolio management is a key resource allocation and balancing activity in many organizations, because the pool of available resources for carrying out projects is generally not sufficient to support the entire pool of projects available for selection. Organizations therefore need to make choices regarding which projects to start, to keep, and which ones to terminate.

The literature demonstrates both financial and non-financial benefits for organizations that apply project portfolio management, such as higher value projects and fewer project delays respectively. However, Jeffery and Leliveld [11] and De Reyck et al. [4]

argue that the benefits of project portfolio management differ between organizations, depending upon the extent to which all project portfolio management practices are in place (a concept known as project portfolio management maturity). Moreover, as Blichfeldt and Eskerod demonstrate, a host of smaller projects are generally carried out 'under the radar' [5]. That is, small projects may not be subject to project portfolio management even in organizations that do have a mature project portfolio management process in place. Hence, firms may experience difficulties in achieving the potential benefits of project portfolio management.

The research described in this paper investigates the approaches that organizations apply to perform project portfolio management. It aims to find out what benefits organizations reap from their project portfolio management implementations, what pitfalls they may encounter and how to avoid these.

We have structured this paper as follows. After the introduction, chapter two provides an overview of project portfolio management, based on a literature study. Chapter three describes our research approach. Chapter four provides the results from fifteen interviews conducted for this research. Chapter five compares the theory on project portfolio management with the practices that surfaced during the interviews, leading to conclusions and further research.

2 Project Portfolio Management

This chapter provides an overview of project portfolio management. Oftentimes, there are more ideas and projects available for selection and execution than the available resources allow for [5, 12,13]. This calls for some form of framework on the basis of which firms can decide whether or not to carry out or terminate projects. Project portfolio management provides such a framework and considers the entire portfolio of projects that a company is engaged in [4]. Archer and Ghasemzadeh base their definition of a project portfolio on the description of projects given by Archibald [14]. A project portfolio would be "a group of projects that are carried out under the sponsorship and/or management of a particular organization" [12].

As a basis for our research, we use the definition provided in [5], stating that *"project portfolio management entails the managerial activities that relate to (1) Screening, selecting and prioritizing of project proposals, (2) Reprioritizing of running projects and (3) Allocating and reallocating resources to projects based on their respective priority."*

The first task would comprise screening, selecting and prioritizing of project proposals based on for instance uncertainty/risk estimations, financial parameters, and resource requirements [12]. The second task would entail reprioritizing running projects based on project status data [2]. Finally, the third task of project portfolio management would encompass that organizations take into account resource constraints and adjust their project portfolio according to the earlier established priorities [13].

In [15] it is argued that projects are only successful if they deliver benefits to the user or owner of the project result. Information on the outcomes of a project would be required to assess whether benefits have indeed been delivered to the project's user or owner. Hence, it is argued here that a fourth task is relevant in addition to the three tasks mentioned above. Practice shows, that tasks 1 and 3, though different in nature, are often combined into one. Therefore, our investigations into project portfolio management has the following tasks in scope: (1) screening, selecting, prioritizing and allocating resources to project proposals, (2) monitoring and reprioritizing running projects, and (3) tracking and managing the realized benefits of projects. The subsequent sections discuss the three tasks of project portfolio management in more detail.

2.1 Screening, Selection, Prioritizing and Allocating Resources

2.1.1 Screening

Screening involves the evaluation of project proposals before projects are selected and added to the project portfolio. Several methods for screening exist. The business strategy method entails that organizations use their strategy to assess which projects to include in their portfolio. These organizations generally distinguish strategic envelopes or strategic buckets to which projects are assigned.

Levine [2] argues that risk should be incorporated into financial project assessments and proposes that risk be incorporated in the form of a discount factor. Archer and Ghasemzadeh [13] and McFarlan [6] focus on the overall portfolio risk level and state that high and low-risk projects should be balanced in the portfolio. This balance would help to prevent that an organization leaves gaps in the market for competitors to fill and it would help to ensure continuance of day-to-day operations. A risk balance in the portfolio of projects would achieve the before mentioned by fostering innovative yet riskier projects that can help build competitive advantage in the future, while at the same time incorporating low-risk projects that support and enhance present-day operations.

The most common method for screening project proposals is the financial method. Relying predominantly on quantitative measures such as financial metrics may result in sub optimal decisions, since crucial qualitative aspects may be overlooked [16] and too much confidence could be placed in the ability of the firm to forecast financial data. Financial screening encompasses some form of profitability or return metric, such as Net Present Value (NPV), Internal Rate of Return (IRR), Return on Investment (ROI), or payback period [4, 11, 13, 16, 17]. Any one of these methods has the potential to be used effectively, yet all have their advantages and disadvantages. Payback period for instance, is a relatively straightforward and easy to explain method, yet it does not take into account any cash flows beyond the payback period [16]. Hence, comparing projects of different duration is complicated. Another example, ROI would be more useful as a performance indicator than as a project evaluation metric, because it does not take into account the time value of money. In order to forecast future cash flows, companies can rely on market research, for instance in the form of consumer panels and focus groups [13].

2.1.2 Selecting Projects and Setting Priorities

Archer & Ghasemzadeh [13] and Cooper et al. [17] state that selection and prioritizing models should be applied consistently so that projects can be equitably compared regardless of the particular model that is used.

Firms may for instance use bubble diagram modeling to select projects and set priorities. Here, projects are plotted on a map using some form of bubbles or balloons. Projects are categorized and resources allocated depending upon what zone or quadrant on the map the projects are assigned to. The axes that are used to create the map can differ and can be for instance risk versus reward, or cost versus timing. Another approach is the scoring model method, in which (potential) projects are evaluated on the basis of a number of ratings or scores that may or may not be weighted to form an overall score for the project. Scoring models are generally used as a ranking or prioritization tool, as opposed to using project scores for go/kill-decisions.

In [18], linear programming is proposed as a model for selecting projects and setting priorities. The model can be used to arrive at a portfolio that is optimized for a certain predefined objective. When this objective is a financial metric such as NPV, the model can optimize for the objective directly. If the objective would be a qualitative measure such as strategic alignment, a quantitative score would first have to be derived. Limited resources should be included in the model as a constraint, as well as other prerequisites such as regulatory compliance projects and running projects that the firm does not wish to interrupt.

Finally, checklists combine criteria by evaluating projects on the basis of a number of yes/no questions. This method is arguably the most straightforward: a project must achieve a designated number of yes-answers in order to be accepted into or remain in the portfolio of active projects. In contrast to the scoring model method, the checklist method tends to be used for making go/kill-decisions rather than setting priorities.

2.1.3 Allocation of Resources

If an enterprise's constrained resources are not allocated effectively, project delays may result because projects have to be put on hold when there are insufficient resources to fund them [2] This phenomenon is referred to as pipeline gridlock: projects keep being added to the list of running projects without taking into consideration resource availability and they are consequently held up as a result of insufficient resources to fund an infinite number of projects. Resources can be allocated by determining the resources available to carry out projects and subsequently assigning those resources to proposed projects according to their relative priority. A first step is to determine what resources are available and whether they suffice to complete currently running projects. Subsequently, firms should take into account proposed new projects and consider whether the available resources allow for starting these projects. This analysis of capacity and demand will demonstrate possible resource shortages. When shortages become apparent, either more resources should be allocated or certain projects should be terminated or reprioritized [13].

2.2 Monitoring and Reprioritizing Running Projects

Once projects are selected, they need to be monitored individually on the project level, and taken together at the portfolio level [2, 13]. Monitoring is important because the environments in which projects operate are not static and projects do not necessarily always run according to plan [11]. The assumptions that were made when the project was started may lose their validity over time, whether expected or not, which may require reprioritizing of projects in the portfolio. Thus, projects need to be periodically assessed in terms of their status and performance [2] Companies that do not reassess their portfolio of projects on a regular basis disregard possibilities that they may have to reprioritize. That is, they forgo possibilities to abandon unpromising projects and to expand investments in successful projects [4]. The current section

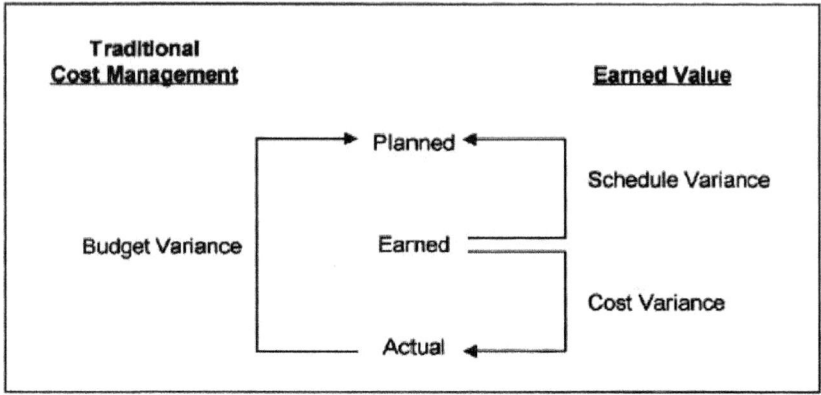

Fig. 1. Traditional cost management vs. EVA [19])

elaborates on three approaches to monitoring projects: earned value analysis, the Stage-Gate process, and the bounding box approach.

Earned value analysis (EVA) is one way to keep track of running projects. Earned value analysis essentially answers the question 'what did I receive for what I spent?'. The difference between traditional forms of cost management and earned value analysis is that the former only compares 'actual cost' with 'planned cost', whereas earned value analysis also incorporates the variable of 'earned value' [19]. As figure 1 shows, earned value analysis disaggregates budget variance into schedule variance and cost variance. It thereby provides insight into the origins of the variance.

Although earned value analysis may provide better insights than traditional forms of cost management, Lukas [20] argues that earned value analysis only works when the organization has reached certain maturity in project management. Earned value analysis requires specific information such as documented project requirements and cost collection systems.

Secondly, organizations may opt for the Stage-Gate process to monitor their running projects, in which projects are divided into phases (each called stages) and decision points (called gates).

The Stage-Gate method breaks down the project process into key activities and decisions as shown in figure 2 [21]. Each stage consists of one or a number of parallel activities that lead up to a subsequent gate. The gates consist of a number of deliverables that decision makers need to make an informed decision for continuance or termination of the project [13]. Gates thus function as a go/kill checkpoint, based on the results of the activities performed in the preceding stage. Figure 2 describes a typical Stage-Gate process for a technological innovation.

A simpler method to monitor and reprioritize running projects is the bounding box approach. This approach can also be used when projects do not fit with the phased Stage-Gate process, for instance when projects are characterized by overlaps between project phases. The bounding box approach entails that the organization sets certain critical parameters (called boundaries) within which the project team itself is authorized to make decisions. Only when exceptions occur will the project portfolio function assess the project and determine whether it should be continued or terminated.

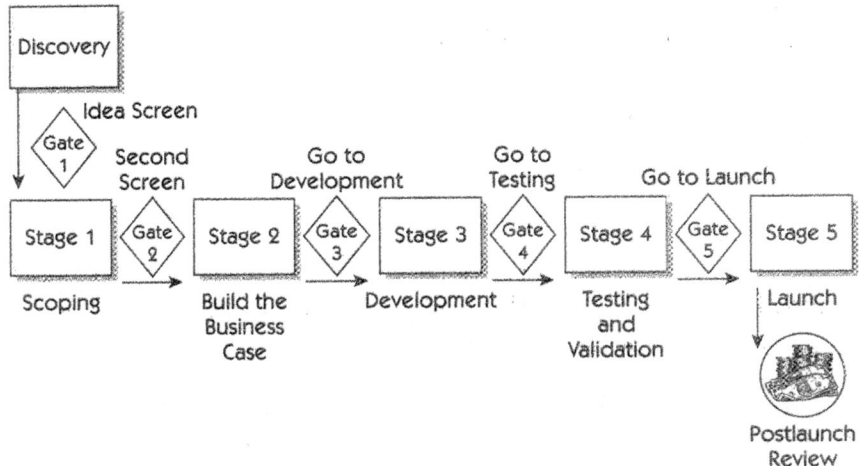

Fig. 2. An overview of the Stage-Gate process [21]

Earned value analysis, the Stage-Gate process, and the bounding box approach are monitoring tools that focus on individual projects and do not consider the entire portfolio of projects. Reviews of the entire project portfolio are needed in addition to the methods that have been discussed in this section so far. The methods that firms apply to screen, select and prioritize project can be used to reprioritize running projects as well [13].

2.3 Benefits Tracking

One would expect firms to learn from their mistakes and success stories to improve their project selection practices. Although Blichfeldt and Eskerod [5] do not mention this issue, other researchers do incorporate benefits tracking as part of project portfolio management in their models [4, 11].

The advantage of tracking the outcomes of projects after completion is that investments in successful strategic buckets can be expanded. Conversely, unsuccessful strategic buckets might require a changed approach or can be scrapped altogether [11]. Without a process to measure the actual benefits of projects however, how would an organization know which ones are successful and which ones are not? Information regarding the success of projects thus becomes a crucial input for the first three tasks of project portfolio management identified in [5]. This feedback concept may be called 'outcome tracking' or 'benefits tracking' and appears to be a necessary component for optimizing the tasks of screening, selecting, prioritizing and reprioritizing, and allocating resources to projects. Without benefits tracking, organizations do not know whether their project investments have been worth the effort or whether they yield a positive return at all. Companies might have trouble implementing benefits tracking because they never set objectives or standards to compare outcomes against. Furthermore, the scope of projects can change over time, rendering the initial standards or objectives invalid and requiring an updated set of standards/objectives [4].

3 Approach

This chapter provides an overview of the methods used to conduct the research. The way the study is set up is discussed, as well as the data collection methods and the sample characteristics. Finally, the chapter discusses the methodology that was used for data-analysis.

3.1 Data Collection Method

Data for the literature overview are collected from secondary sources, such as academic and practitioners' journals, books, and published Websites. Primary data are subsequently collected through semi-structured interviews. The interviews are guided by a predefined topic list, but there is room for deviation and variation depending on the flow of the interview. The interviews are set up in a semi-structured way to ensure that meaningful responses can be elicited from the respondents depending on their knowledge and the organizational context. The interviews are held with respondents at client firms of Ernst & Young to discover what their project portfolio management practices are and how they – if at all – benefit from these practices. In a majority of interviews, two interviewers are present during the interview to enhance the flow of the conversation and to ensure that all applicable topics are covered. A drawback of this approach is the sometimes erratic course of the interviews. To ensure that non-verbal cues can enrich the data, the interviews are conducted on-site and face-to-face where possible. The interviews are conducted by phone in a minority of instances, where an on-site appointment was not possible. The interviews are audio recorded with the respondents' permission and subsequently transcribed.

3.2 Sample

Fifteen respondents with knowledge of project portfolio management within their respective organizations were identified for the empirical study. The aim has been to find organizations that are aware of project portfolio management so that they can provide insights on how project portfolio management can be beneficial. Saunders et al. [22] refer to this type of sampling as 'purposive sampling', meaning that the judgment of the researcher was used to select respondents who would enable answering the research questions. Advantageous about the purposive sampling strategy is that people knowledgeable in the field of project portfolio management could be identified efficiently. The sample consists of fifteen respondents, representing thirteen companies.

3.3 Data-Analysis Procedure

The transcript data have been categorized using the interview topic list. The data display consists of all topics and all literal responses extracted from the transcripts that pertain to the topic in question. Subsequently, similar responses have been sought within categories and these were counted and grouped. This analysis forms the basis of chapter four, as displayed in graphs 1-9.

3.4 Survey

On top of the 15 real-life cases, we have validated the results with an online survey among 650 respondents. The results of this survey are currently analyzed and will be addressed in a future paper. First results show no substantial deviation from the findings of the interviews.

4 Findings

This chapter starts by characterizing the project types that respondents consider in their project portfolio management and it subsequently elaborates on the four core tasks of project portfolio management. We conclude by discussing the respondents' view on the critical success factors for project portfolio management, their view on the advantages of and pitfalls for project portfolio management, and the improvement areas that interviewees have identified within their respective organizations.

4.1 Project Types

Eight of fifteen respondents reported that their role in project portfolio management is limited to projects that involve IT. One respondent was concerned with a portfolio of new product development (NPD) projects. A group of four respondents reported that their project portfolios consider projects of all types. Finally, two respondents indicate that their project portfolios contain mainly infrastructure projects.

Respondents who focus on IT provide two reasons for this emphasis. First, five respondents state that virtually any project involves IT to some extent, because business processes generally depend on a certain IT-infrastructure. Making changes in the organization hence quickly leads to changes in the underlying IT-infrastructure. Second, two respondents state that employees outside the IT discipline are unfamiliar with keeping track of time spent on projects for reporting purposes and that these records are needed for assessing the status of projects. Unfamiliarity with timekeeping would therefore increase complexity of introducing project portfolio management for non-IT projects.

4.2 Project Screening and Selection

4.2.1 Financial and Strategic Screening

The topic of financial metrics has been discussed with twelve respondents, all of whom report that their firms rely on multiple financial metrics for project screening. Within the plethora of financial metrics, the most common metric reported by the respondents was the net present value (NPV) measure, which is reported by eight respondents. The pay back period (PBP) method comes in second at seven mentionings. Return on investment (ROI) was reported by five respondents; internal rate of return (IRR) by four, and the absolute cost of the project has been mentioned by two respondents as a financial measure for screening and selecting projects and for determining their relative priorities.

Although three respondents said they regard the ease of use of the payback period method as beneficial, most respondents did not provide a substantive rationale for the choice of particular financial metrics. Another three respondents indicated they do not see differences between various financial metrics and two respondents indicated they do not know why their firm opts for particular financial metrics.

Furthermore, ten respondents reported that they consider alignment with company strategy when screening project proposals. In seven cases strategic alignment was used as part of a set of multiple criteria for screening, selection, and prioritizing. Two approaches to considering strategic alignment can be discerned. The first approach is to determine strategic objectives and to determine what actions are needed to achieve these objectives. Projects are then derived from each of or combinations of these actions. The second approach is to allow employees to propose projects as the need for change arises. These proposals are then screened to verify whether the proposed changes fit with strategic objectives.

4.2.2 Risk Analysis in Screening

Eleven of fourteen respondents reported that their companies consistently apply a risk metric to screen projects. Risk metrics that respondents mentioned vary and include feasibility, complexity, and market dominance. Furthermore, three respondents reported that their companies link project risks and returns to each other. As opposed to considering risk as a separate item, these three firms link the degree of certainty with which a project can be completed successfully to other elements of the project score. More specifically, one interviewee reported that his company makes risk adjustments to the *overall* project score, meaning that projects for which the expected risks are high receive a lower score. The two others in this group adjust their *financial metrics* according to the anticipated project risks. For example, cost expectations for a project may be doubled if the technology risk is considered to be high due to the introduction of a new type of technology.

4.2.3 Selection and Prioritizing

Two respondents indicated that they use the scoring model method, where each project proposal receives a score based on multiple criteria and projects are prioritized according to their relative scores. Twelve others reported that they do consider multiple criteria for selection, but that they do not combine these criteria into a selection model. Selection and prioritizing criteria that are used for both approaches are financial metrics, the feasibility of the project, the extent to which the proposed project is in line with the company's strategy to ensure projects contribute to achieving strategic objectives, and compliance with government laws and regulations to prevent fines and other governmental reprimands. Some criteria may be industry specific and hence not applicable to organizations in general. Examples in this category include customer safety and environmental impact.

4.3 Monitoring Current Projects

All respondents indicated that they have a centralized idea of what projects are currently running and what the status of those projects is. Three respondents reported that their organizations use the Stage-Gate process to monitor running projects. Five respondents stated that their firms use a form of the bounding box approach, where the project team is authorized to make its own decisions within certain boundaries. Only exceptions that are outside these boundaries are reported to bodies that are higher in the organizational hierarchy, such as an investment board or portfolio board. Seven respondents reported that their firms use a fully centralized approach to project control. Here, all projects report to a body that is higher in the organizational hierarchy than the project team itself on a regular basis.

As discussed in chapter two, escalation of commitment occurs when current projects are not adequately monitored, resulting in unwarranted continuation of unpromising projects. Two respondents explicitly indicated they had not experienced escalation of commitment as a real issue. They argued that proper project screening and selection practices largely prevent the occurrence of unpromising projects altogether by ensuring that only promising projects are carried out. Furthermore, adequate monitoring would avert derailment of projects.

4.4 Internal Constraints and Allocating Resources

Fourteen of fifteen respondents were able to provide information about their respective companies' approach to constraints that are internal to the company and that influence the allocation of resources to projects. Two types of constraints surfaced during the interviews: scarce resources and sequential dependencies between projects. Ten respondents reported that they only consider resource availability. A total of four respondents indicated that internal constraints need to be more carefully considered in future. Another two respondents indicated that they actively consider whether projects do not interfere with each other, in terms of both resources and sequential dependencies. One of these respondents reported that certain meetings are dedicated to consider project interdependencies and the other respondent indicated that company wide requirements are collected early in the project portfolio management cycle. The latter respondent said that creators of a project proposal are asked to consult all stakeholders within the company to verify whether the proposed project may influence running projects or existing IT platforms (the firm applied project portfolio management to its portfolio of IT projects).

In total, thirteen respondents reported that resource constraints are taken into account at some point in the project portfolio management process. The types of resources that surfaced during the interviews are financial resources and human resources. Eleven of thirteen respondents indicated that they take into account both financial and human resources in their project planning to enable cost control and to ensure that projects can be adequately staffed.

4.5 Benefits Tracking

Nine respondents reported that the benefits of completed projects are not tracked at all or that project benefits are tracked on an ad-hoc basis in incidental cases. One interviewee reported that the concept of benefits tracking is not applicable to all projects because certain projects are imposed upon the organization. Only one respondent reported that the organizational body that is responsible for project portfolio management consistently tracks the benefits of completed projects. His firm measures the actual outcomes of projects three months after they have been closed and, when deemed necessary, a year after their closure. The respondent explains these so-called 'post calculations':

"We mainly look at the benefits because there are projects that introduce a certain new service to the market for which we really want to know what their return is and whether it was worth the effort. There are also projects that for instance replace a certain system, maintenance on projects or licenses for ERP systems, etc. Those projects don't return anything and so it does not make sense to do post calculations. In those cases we do of course check whether they have remained within budget."

Another two respondents indicated that they have recently engaged in benefits tracking and that the benefits of the projects that are currently running will be tracked once closed. Finally, two respondents reported that their respective organizations do track the benefits of closed projects, but that this responsibility is delegated to other parts of the organization. As a consequence, there is no feedback mechanism from the actual outcomes of projects back to the criteria that are used for projects screening, selection, and prioritizing. For instance, if some organizational subunits consistently perform better than others, this difference in performance would not be reflected by the allocation of resources across subunits.

Why have most organizations in the sample not engaged in benefits tracking so far? Five respondents indicated that it is complicated to attribute benefits to the right causes. The actual benefits of a project, such as a cost reduction or increased revenues, may be caused by factors other than the particular project. The attribution problem may be enhanced when benefits are due to take effect and thus measured a considerable period of time after the project is closed. A potential solution that has been proposed is to incorporate the anticipated benefits of a project into the first upcoming budgeting cycle so that the project and its anticipated effect are as close as possible to each other in terms of time.

4.6 Critical Success Factors

Central to the success of any project portfolio management implementation would be the commitment of top management, according to seven of fifteen respondents. This commitment would be necessary to ensure that all organizational bodies and individuals that are affected by the project portfolio management framework are either convinced of or forced to commit to its cause.

Five respondents reported that one critical success factor is the realization that project portfolio management requires a pragmatic approach to be successful. Interviewees indicated that the project portfolio management process should be formalized and rigid to the extent that the organization carries out projects that are started and monitored on the basis of predefined and objective criteria. However they also stated that the process should be flexible enough to allow for speedy decision-making. Respondents indicated that organizations should therefore be critical of the amount of regulations and templates they impose on project managers and sponsors and that this administrative burden should be kept to a minimum.

Five respondents stated that another critical success factor is transparency regarding the grounds on which decisions are made, and regarding the status of running projects. Transparency on these issues would facilitate learning about the organization's strengths and weaknesses in project management, in addition to creating awareness for the importance project portfolio management and increasing acceptance for project portfolio management within the organization.

Finally, two respondents advised to keep the project portfolio management function small. One of these respondents referred to the number of tasks of the project portfolio management function and urged to start off with a limited and hence easier to handle number of tasks. The second of these respondents referred to the number of employees within the project portfolio management function. The respondent argued that keeping the function small leads to more networking and communication within the organization because time pressure would force employees to look for innovative ways to handle their workload.

4.7 Project Portfolio Management Advantages and Pitfalls

4.7.1 Advantages of Project Portfolio Management

Respondents stated that the project portfolio management process enabled them to make the right decisions for the right reasons. Project portfolio management provided them with means to prevent opportunism in starting projects. Project portfolio management would help to ensure that project proposals are assessed based on objective criteria so that only useful projects are started. The project portfolio management process apparently forces firms to answer questions such as 'does this project have a solid business case?', 'do we have budget to carry out this project?', 'does it fit our planning?' and 'do we have sufficient human resources available?'

Project portfolio management would also enable both control over and reduction of costs by stopping projects that are not likely to generate positive returns and by not starting unpromising ones to begin with. Furthermore, centrally tracking active projects and project proposals appears to enable organizations to prevent budget overruns on the project portfolio as a whole. One respondent was particularly explicit about the cost saving benefits of project portfolio management in the short term. He argued that stopping redundant projects and thereby saving money is a quick win of project portfolio management:

"When we started with [project portfolio management] towards the end of 2007 we had 400+ running projects and now (...) that's close to 100. (...) As soon as you start inventorying what you have across your entire group you'll encounter easy wins. (...) Project costs to start with."

Project portfolio management would also enable managing the overall value of their project portfolio. Project portfolio management would help to achieve strategic objectives and to keep track of whether intended project benefits are eventually achieved. In addition, firms now know what projects are running and what the status of these projects is as a result of their project portfolio management practices. One respondent summarized the above and said that the advantages of project portfolio management are:

"A general idea of what's running, what's coming and what the status of everything is. Plus, not completely unimportant is to link that to what [the project] costs, whether we actually want that and what we aim to achieve with [the project]."

Synergies between projects can more easily be identified because project portfolio management would prescribe centrally tracking all projects, thereby enabling the identification similarities and overlap and subsequently enabling corrective action. Finally, one respondent said that project portfolio management helped to achieve a balanced project portfolio in terms of discretionary versus obligatory projects.

4.7.2 Pitfalls for Project Portfolio Management

Respondents indicated that a perception of bureaucracy that spurs uncooperative behavior at lower management levels is a pitfall for project portfolio management. The resulting resistance on the part of employees may undermine the objective of project portfolio management to achieve strategic objectives, because ultimately people have to make projects happen. If these people resist the methodology they may be less effective at carrying out the project. Hence, the process of achieving strategic objectives and moving forward may be stifled. Resistance may also be caused by decreased freedom (noted by three respondents) for employees and increased transparency (noted by two respondents) about the reasons for starting a project and the status of running projects.

Furthermore, the project portfolio management function may impose the filing of forms, usage of document templates, and may simply require more administrative operations than would be necessary from the project manager's or projects sponsor's point of view. One respondent provided an illustrative example:

"Do you know how that goes? 'I have received your files, but you should have handed them in on Wednesday for next week's executive meeting and that's cramped already. And we have things that are so important right now... your turn will be next time.' That sets you back another two weeks. And then you've just missed the [for instance budgeting] cycle that is once a month... if you're out of luck you'll be delayed for a couple months."

Although the respondents mentioned the pitfalls discussed here, the first response of four respondents was that they did not see any pitfalls. The advantages of project portfolio management appear to strongly outweigh the pitfalls for these respondents.

One pitfall for project portfolio management described in chapter 2 is the phenomenon that a host of smaller projects that operate under management's radar might undermine the effectiveness of the project portfolio management process. Eight respondents reported that their firm uses a cut-off budget below which project proposals are handled differently. Six respondents explained their alternative methods towards smaller projects. Five of these entailed delegation of the responsibility to lower level management. The sixth respondent indicated that smaller projects are discussed in roundtable meetings where business and IT representatives are present to discuss what needs to be done.

Several project size cut-offs are mentioned ranging from 50.000 euros (two respondents) via 100.000 euros (two respondents) and 200.000 euros (one respondent) to 2M euros (one respondent). Another two respondents indicated that their respective firms do use a project size cut-off but they did not know the exact amount.

4.8 Improvement Areas

Seven respondents indicated that project benefits are not consistently tracked and view benefits tracking as an improvement area for their respective organizations. Two of these respondents stated that the rationale behind their desire to implement benefits tracking is to learn from past mistakes in an attempt to improve future performance. The following quote illustrates this rationale for benefits tracking:

"Business cases are prepared, the project is carried out, everyone is happy, customers are using their new services... And then no one actually looks back whether it went better or worse compared to the business case. That's where you miss out on the learning curve."

One respondent who views benefits tracking as an improvement area highlighted that benefits tracking is not a prerequisite for successful projects. The respondent explained that projects might be successful, but that benefits tracking enables the organization to identify and assess mistakes. This knowledge can then be used to prevent the same mistakes in future or to replicate successful practices improve future practices, thereby improving future performance.

Six respondents reported that project portfolio management within their respective organizations should be done more efficiently and involve less bureaucracy. These respondents wanted to increase decision-making speed by for instance reducing the bureaucratic burden imposed on project teams and through better software support.

Three respondents indicated that they would like more insight in the portfolio of projects in general. That is, they value a more integrated, holistic idea of what the overall status of the project portfolio is and better mechanisms for identifying possible performance deficiencies. This would make it easier for them to take corrective action when and where needed.

5 Conclusions, Discussion and Further Research

This chapter discusses the similarities and differences between the theory on project portfolio management and the results found in the interviews. The first three sections below each discuss and conclude on one of the research questions. The answer to the problem statement is formulated in the recommendations section. Finally, some suggestions for future research are outlined.

5.1 Comparing Research Outcomes with Existing Literature

5.1.1 Screening, Selecting and Prioritizing Project Proposals
All interviewees indicated that they apply a selection of the financial methods proposed in literature to screen project proposals. However, literature also states that each of the financial methods has its advantages and disadvantages, and that their

effectiveness depends on the way they are put to use; for instance whether they are used as criteria for project screening or as performance indicators once the project is carried out. This suggests that a conscious choice of financial methods is in order. Nevertheless, the interview data indicate that choices made with regard to financial methods are largely not consciously made.

Literature found that the best performing project portfolios are governed by multiple methods for screening, selecting and prioritizing project proposals. Only a limited number of interviewees indeed apply multiple methods for screening, selection and prioritizing. Even though a combination of screening methods is often applied, these methods are rarely combined into selection and prioritizing models. Only a few of the respondents combine their screening methods into a model that considers the relative importance of the methods.

5.1.2 Monitoring and Reprioritizing Running Projects

With regard to the three approaches to monitor running projects proposed in literature, none of the interviewees reported the application of earned value analysis. A third of them did report the application of the bounding box approach. This could be an indication that firms would rather opt for a simpler approach, since the bounding box approach requires less mature project management practices. Perhaps the project management practices in the organizations that the interviewees work for do not have the information systems (e.g. documented project requirements and cost collection systems) in place to enable earned value analysis. All interviewees do report that they have a process to monitor the status of running projects. Hence, they do not run the risk of disregarding possibilities to abandon unpromising projects and to expand their investment in successful projects [4]. Organizations may opt for any of the available project control mechanisms, as long as they consistently apply them to all projects. This ensures that projects are comparable and that firms can make informed decisions for continuing, terminating, or correcting projects.

5.1.3 Benefits Tracking and Management

Most interviewees reported that their respective companies do not track realized project outcomes. This finding is consistent with earlier studies, even though benefits tracking is important and has substantial potential advantages as discussed in chapter two. One of the problems associated with benefits tracking is the complexity of attributing benefits to individual projects. The solution proposed earlier is to feed forward project outcomes into budgeting cycles. This way, benefits are automatically taken into account at the aggregate level. Then, corrective action can be taken when needed. For instance, if the overall return on investment for projects is low, an organization might want to change the criteria on the basis of which it accepts projects. Another possible solution would be to assess the feasibility of benefits in project proposals in such a way that there is less clutter that could moderate or mediate the relationship between project efforts and realized benefits. Finally, if a project is closed and benefits tracking indicates the anticipated benefits have not been achieved, the organization could commence a new project in an attempt to achieve the intended benefits still. Actively following up on planned but non-realized benefits is referred to here as benefits management.

5.2 Project Portfolio Management Advantages and Pitfalls

5.2.1 Advantages Associated with Project Portfolio Management

The most noted advantages found during the interviews are 'making the right decisions for the right reasons,' 'enabling cost control and reduction,' and 'managing the overall value of the project portfolio.' These advantages correspond to the value creation advantage identified in the literature [1, 2, 4, 11], where selecting the most promising projects and cost savings can create value for the organization. In addition, the advantage of project portfolio management noted by one interviewee is 'the ability to balance the portfolio' and this topic also recurs in the literature discussion on value creation.

The second advantage identified in the literature is the ability to manage uncertainty and risk [6, 13, 23]. The interviewees did not mention this advantage as such. Existing literature explains the ability to manage uncertainty and risk as an example of a learning organization and perhaps this requires a maturity level that the organizations that the respondents work for have not yet reached. As discussed in the sections on benefits tracking, organizational learning is a concept that respondents are aware of but it is also a concept that has not fully come to fruition yet.

The ability of organizations to terminate undue projects [2, 24, 25], is not mentioned as an advantage by the interviewees. Rather, respondents point at the ability of project portfolio management to enable selecting and monitoring projects on objective grounds as a beneficial factor that decreases occurrence of unpromising and derailed projects.

5.2.2 Pitfalls Associated with Project Portfolio Management

The first pitfall for project portfolio management is generating sufficient buy-in from all organizational levels [11, 26, 27]. Although assembling adequate buy-in from all organizational levels was not reported as a pitfall for project portfolio management by the interviewees, the buy-in from top management is the most frequently mentioned critical success factor. Moreover, a perception of bureaucracy is the most noted pitfall for project portfolio management by the interviewees. This perception may originate from a lack of buy-in, since employees may see the project portfolio management function as taking away their flexibility, freedom, and independence. Organizations should make sure that their employees understand the reasons for the implementation of project portfolio management and they should demonstrate how project portfolio management can be helpful rather than detrimental to them. Firms can for instance demonstrate how project portfolio management can solve some of the issues that employees experience in their daily project-related work.

A second pitfall are the difficulties to find the time and information required for project portfolio management [8, 26, 28]. Several interviewees also contended that the additional time that it takes to start projects because of project portfolio management is a pitfall for the process. Speeding up the process of approving project proposals by, for instance, increasing the frequency project proposal review meetings may alleviate this challenge.

Furthermore, in [29] it is stated that it is often difficult to make accurate estimations of the outcomes of project parameters and that firms should therefore not rely too heavily on quantitative selection criteria. Notably, none of the interviewees have reported the limited ability to estimate project outcomes as a pitfall for the project portfolio management process. This may be due to the fact that the literature highlights the challenge of estimating project outcomes in the context of purely quantitative methods for project selection such as linear programming. None of the respondents indicated that they used linear programming for project selection, or that they rely merely on quantitative methods. It therefore appears that respondents alleviate the challenge of accurately estimating project outcomes by relying on multiple methods for project selection.

Regarding the issue of un-enacted projects, it is interesting to observe that none of the respondents reported that they regard smaller projects as a problem. In other words, none of the respondents support Blichfeldt and Eskerod's notion that the value of project portfolio management is endangered by un-enacted projects [5]. It appears that the companies the respondents work for have come up with solutions to the un-enacted projects pitfall: three respondents reported that they have a separate budget for smaller projects, which enables them to control the costs of these initiatives. Four respondents indicated that there are guidelines and criteria for small projects to enable monitoring and control of smaller initiatives.

5.3 Conclusions

To benefit from project portfolio management, organizations should consistently consider multiple methods for screening, selection and prioritizing that are widely supported by key stakeholders. The organization in its entirety should be made aware of the function and benefits of project portfolio management, for instance by demonstrating how project portfolio management can resolve project-related issues that employees encounter. Creating widespread awareness and support within the organization is important for the proper functioning of project portfolio management.

Secondly, it is essential to find a balance between qualitative and quantitative methods for screening, selection, prioritizing and resource allocation, because over-reliance on quantitative methods entails the risk of overestimating the organization's ability to accurately approximate project outcomes. The majority of firms in the sample can improve by combining financial and business strategy screening methods into a scoring model that takes into account the relative weights of the methods and by considering bubble diagrams. Furthermore, most respondents did not provide a substantive rationale for the choice of particular financial metrics. Firms should make deliberate choices in this regard, because the effectiveness of each of the financial metrics depends on the way they are applied.

Organizations should track and manage project benefits because realizing those benefits is the primary objective investments in projects. Organizations can identify opportunities for improving their screening, selection, prioritizing and resource allocation processes through the application of benefits tracking and they can attempt to still realize project outcomes that were initially not achieved through benefits management. Finally, organizations need to ask themselves which project portfolio management elements add value and which ones do not. By eliminating elements that do not add value, the administrative burden required for project portfolio management is minimized in an attempt to increase decision-making speed and flexibility.

5.4 Suggestions for Future Research

A larger empirical study would be required to link project portfolio management practices to financial performance to quantify the value of project portfolio management. We have started this by conducting a survey among 650 respondents. Research across multiple industries and countries could verify whether the results found in this research apply to a wider range of businesses. Future research should preferably involve multiple interviews with each respondent and with multiple respondents within the same organization. This approach can shed light on possible differences in how project portfolio management is perceived by varying stakeholders within organizations. Finally, respondents hinted at the interactions between project management and project portfolio management. This topic has not been discussed in the literature before. A key question here would be how mature project management practices within an organization should be to implement project portfolio management successfully or the other way around.

References

1. Dougherty, R.: Section 9.3: Developing a project portfolio management Capability at America Online. In: Levine, H.A. (ed.) Project Portfolio Management - A Practical Guide to Selecting Projects, Managing Portfolios, and Maximizing Benefits. Jossey-Bass, San Francisco (2005)
2. Levine, H.A.: Project Portfolio Management - A Practical Guide to Selecting Projects, Managing Portfolios, and Maximizing Benefits. Jossey-Bass, San Francisco (2005)
3. Markowitz, H.: Portfolio Selection. Journal of Finance 7(1), 77–91 (1952)
4. De Reyck, B., Grushka-Cockayne, Y., Lockett, M., Calderini, S.R., Moura, M., Sloper, A.: The impact of project portfolio management on information technology projects. International Journal of Project Management 23, 524–537 (2005)
5. Blichfeldt, B.S., Eskerod, P.: Project portfolio management – There's more to it than what management enacts. International Journal of Project Management 26, 357–365 (2008)
6. McFarlan, F.W.: Portfolio approach to information systems. Harvard Business Review 59(5), 142–150 (1981)
7. Cooper, R.G., Edgett, S.J., Kleinschmidt, E.J.: New Product Portfolio Management: Practices and Performance. Journal of Product Innovation Management 16, 333–351 (1999)
8. Killen, C.P., Hunt, R.A., Kleinschmidt, E.J.: Project portfolio management for product innovation. International Journal of Quality and Reliability Management 25(1), 24–38 (2008)
9. Jeffery, M., Leliveld, I.: Best Practices in IT Portfolio Management. MIT Sloan Management Review 45(3), 41–49 (2004)
10. Archer, N.P., Ghasemzadeh, F.: An integrated framework for project portfolio selection. International Journal of Project Management 17(4), 207–216 (1999)
11. Cooper, R.G., Edgett, S.J., Kleinschmidt, E.J.: New Problems, New Solutions: Making Portfolio Management More Effective. Research Technology Management 43(2), 18–33 (2000)
12. Archibald, R.D.: Managing High-Technology Programs and Projects. Wiley, New York (1992)

13. Turner, J.R., Cochrane, R.A.: Goals-and-methods matrix: coping with ill defined goals and/or methods of achieving them. International Journal of Project Management 11(2), 93–102 (1993)
14. Remer, D.S., Stokdyk, S.B., Van Driel, M.: Survey of project evaluation techniques currently used in industry. International Journal of Production Economics 32, 103–115 (1999)
15. Cooper, R.G., Edgett, S.J., Kleinschmidt, E.J.: Portfolio management for new product development: results of an industry practices study. R&D Management 31(4), 361–380 (2001)
16. Chen, M.T.: The ABCs of Earned Value Application. AACE International Transactions (2008)
17. Lukas, J.A.: Earned Value Analysis - Why it Doesn't Work. AACE International Transactions (2008)
18. Cooper, R.G.: Section 7: project portfolio management applications: new product development. In: Levine, H.A. (ed.) Project Portfolio Management - A Practical Guide to Selecting Projects, Managing Portfolios, and Maximizing Benefits. Jossey-Bass, San Francisco (2005)
19. Saunders, M., Lewis, P., Thornhill, A.: Research Methods for Business Students. Pearson Education, Essex (2007)
20. Olsson, R.: Risk management in a multi-project environment: An approach to manage portfolio risks. International Journal of Quality and Reliability Management 25(1), 60–71 (2008)
21. Keil, M., Mann, J., Rai, A.: Why Software Projects Escalate: An Empirical Analysis and Test of Four Theoretical Models. MIS Quarterly 24(4), 631–664 (2000)
22. Harrison, P.D., Harrell, A.: Impact of Adverse Selection on Managers' Project Evaluation Decisions. The Academy of Management Journal 36(3), 635–643 (1993)
23. Datz, T.: How to Do It Right. CIO 16(14) (2003)
24. Cohen, C.B., Englund, R.L.: Section 5.1: Making the Case for Project Portfolio Management. In: Levine, H.A. (ed.) Project Portfolio Management - A Practical Guide to Selecting Projects, Managing Portfolios, and Maximizing Benefits. Jossey-Bass, San Francisco (2005)
25. Morcos, M.S.: Modelling resource allocation of R&D project portfolios using a multi-criteria decision-making methodology. International Journal of Quality and Reliability Management 25(1), 72–86 (2008)
26. Ghasemzadeh, F., Archer, N.P., Iyogun, P.: A Zero-One Model for Project Portfolio Selection and Scheduling. The Journal of the Operational Research Society 50(7), 745–755 (1999)

Systemic IT Project Management: A Rational Way to Manage Irrationalities in IT Projects?

Andreas Drechsler, Peter Kalvelage, and Tobias Trepper

University of Duisburg-Essen, Information Systems for Production and
Operations Management, Universitätsstraße 9, 45141 Essen, Germany
{andreas.drechsler,peter.kalvelage}@icb.uni-due.de,
tobias.trepper@icb.uni-due.de

Abstract. Various studies of IT project management work have shown that common causes of project failures belong to the category of soft factors like communication or cultural differences. Instead of combating these causes with the application of even greater numbers of formal project management methods, this paper suggests a different approach derived from the systemic theory of the German sociologist Niklas Luhmann. Projects are viewed and treated as social constructs. The approach is based on the acceptance of the irrationalities in the project environment and attempts to counteract them by understanding and handling them at their true roots instead of only treating the superficial symptoms. Additionally, this perspective leads to a new "mindset" for a project manager who is facing complexity, uncertainty and irrationalities in IT projects.

Keywords: Systemic, IT, project, management, methods, approaches, social, systems, theory, rationality, irrationality.

1 Introduction

The usual way of transforming enterprises is doing so by means of a project. Due to the extensive reliance on IT in business processes in modern organizations, these projects often involve the transformation of the enterprise's IT as well. But project management, and specifically IT project management, nowadays still proves to be a big challenge for project managers. The annual CHAOS report issued by the Standish Group, published since 1994, shows in its 2009 edition ([6] referring to [1]) that only 32% of surveyed IT projects were finished successfully, 44% were delayed or exceeded budget and 24% were cancelled or abandoned entirely. Since its inception in 1994 [2] these numbers have improved somewhat (1994: about 16% successful, 53% delayed / over budget, 31% failed), but still about two-thirds of the surveyed projects cannot be labelled a "complete success".

To help project managers cope with this challenge, there are lots of books, articles, courses and certifications available on the market. International industry standards like PRINCE2 or PMBOK even attempt to provide comprehensive frameworks for successful project management. The general focus lies in the area of formal management methods, to aid with planning, (formal) organization, especially in the area of

F. Harmsen et al. (Eds.): PRET 2010, LNBIP 69, pp. 127–155, 2010.

process organization, or the measurement of success. But still, despite all these efforts, about two-thirds of all IT projects are regarded as unsuccessful in one way or another.

Soft factors like communication, the project culture and the level of trust among the stakeholders are often identified as either main causes for those failures or as major success factors for IT projects (e. g. in [3]), and have become generally acknowledged and represented in project management literature and standards. Yet there are few actual approaches or frameworks for IT projects available who attempt to understand and integrate major soft factors into the general approach to project management right from the onset. In harsh contrast, the commonly cited hard factors like gathering sufficient requirements are often represented as a separate phase or activity throughout the project (as, for example, in the IBM Rational Unified Process [4]). The same applies to generic project management frameworks like PRINCE2 or PMBoK which focus on activities like planning and controlling, or issues like cost, time, quality or risk [47].

The authors hypothesize that by anchoring the soft factors in a project management framework as prominently as the hard factors are anchored in modern project management frameworks, the chance of project failure because of the soft factors will decrease and project success chances will subsequently increase. The approach described here is therefore not to be seen as a replacement for traditional methods of project management, but as a supplement in a vital, and often overlooked area, contributing to successful project management and project completion.

Chapter 2 of this paper will tackle the role and the nature of soft factors or irrationalities in IT projects in greater detail. Due to their social and therefore potentially irrational nature, a different approach to their handling is necessary, compared to the usual handling of the hard factors. The term irrationality in this paper is used to emphasize the element of surprise (pleasant and unpleasant), misunderstandings, unexpected occurrences or the impossibility to anticipate them in a structured and reliable way. As subsequent chapters will show, the theory of social systems by the German sociologist Niklas Luhmann is a suitable theory for establishing recommendations for coping with those irrationalities in a rational way. The necessary fundamentals of his theory are then outlined in chapter 3.

Chapter 4 will subsequently put the theory into practice, first by identifying three major sources for irrationalities in IT projects based on Luhmann's theory (section 4.1), and then by outlining general principles of systemic project management to cope with these irrationalities (section 4.2). Section 4.3 shifts the focus from functional or managerial principles to the institutional perspective of the IT project manager and the consequences for them and their "mind-set" to successfully apply the systemic principles to their daily business of project management. Section 4.4 will conclude the discussion by showing possible barriers to the application of the systemic approach in a project environment in practice.

To illustrate how the general principles can be realized in practice, chapter 5 provides three examples of existing systemic IT project management frameworks and discusses advantages and disadvantages of each method in light of the general principles discussed beforehand.

Due to the high level of abstraction most of the argumentation relies on, the generic, underlying ideas can also be applied in a similar way to the area of enterprise transformation, specifically to the field of organizational change management. This will be exemplified in chapter 6. Chapter 7 will give a conclusion and an outlook for further research.

2 Soft Factors in IT Project Management

This chapter will give an overview of the aforementioned soft factors in IT project management and provide a rationale for the applicability of Luhmann's theory of social systems to the area of IT project management.

There are a sizable number of studies available which analyze success and failures of projects. The results all vary in the particulars, yet the bottom lines are rather similar in general. Two exemplary studies are discussed below.

2.1 Critical Success Factors for IT Projects According to Kotulla

Kotulla [32] conducted an empirical study in an international software development company about critical factors for success and failure. He divides his findings in three categories – managerial, technological and social factors. The factors identified by him are as follows:

- Technical factors: communication infrastructure, specifications, software architecture, unambiguity of requirements, technical suitability, tests, tools
- Managerial factors: correct estimation, project management, priorities, project sizing, shared idea of goals, trust, politics
- Social factors: communication, cultural differences, trust, soft skills

All technical factors except the unambiguous requirements can be counted as hard factors for project success and are commonly represented in typical frameworks and methods for software development. The same applies for the managerial factor of correct estimation of effort. The managerial factor of project management bears the responsibility of ensuring that all other success factors are applied in the best possible way and leading the project to success. Three of the remaining managerial factors (shared idea of goals, trust and politics) as well as all of the social factors can be labelled soft factors for project success.

2.2 Critical Factors for Project Success or Failure by GPM e. V.

A study conducted by the German Association for Project Management (GPM e. V.) in 2008 analyzed both successful projects in terms of success factors as well as failed projects in terms of factors contributing to the failure [31]. As top three success factors they identified qualified project team members, good communication and clear goals and requirements. The top three reasons of failure consisted of bad communication, ambiguous goals and requirements as well as politics, like department egoism and conflicts of competences.

It is striking that two factors appear both as factors for success and failure (communication as well as clear goals and requirements). It is also of note that communication is a major factor influencing the clarity of goals and requirements, so these two factors can be considered linked. With this in mind, only one of the top success factors (qualified team members) can be regarded as somewhat of a hard factor while the five other major factors for success and failure definitely fall into the category of soft factor. With the exception of the qualification of the team members, the factors identified here were also identified by Kotulla, so this study is complementary by providing a more recent analysis as well as a ranking of the top factors for both success and failure across the entire industry.

2.3 Applicability of Luhmann's Theory of Social Systems to the Realm of IT Project Success and Failure

Both studies identify communication as well shared and unambiguous goals and requirements as success factors for projects. As we will see in chapters 3 and 4, ambiguity/uncertainty and communication are central concepts in Luhmann's theory of social systems. Furthermore, cultural differences are expressed and observed via means of communication and soft skills influence the capability of effective, unambiguous communication. Therefore these factors can be regarded as sub factors or side factors to the element of unambiguous communication.

Trust and politics (like department egoisms) were also identified as two important soft factors by both studies and can be interpreted as two side-effects of the established social order within an enterprise, another element of Luhmann's theory as outlined in chapter 3. Therefore it can be said that Luhmann's theory of social systems addresses major factors for success and failure in projects and is suitable for further analysis and discussion in this paper.

Kotulla identified "project management" as separate success factor – chapters 4 and 5 will discuss principles and approaches for the project management to cope with these soft factors or irrationalities in a superior way compared to dealing with them by "best effort" on a case-by-case basis, and well outside their established formal project management framework. By providing both a classification of sources for these irrationalities as well as suggestions how to integrate methods to cope with these problems into a process structure of a project, IT project managers are enabled to handle the irrationalities in a somewhat structured – or rational – way.

3 Luhmann's Theory of Social Systems

The main source for the theoretical foundation of the approach presented in this paper is the theory of social systems by the German sociologist Niklas Luhmann [5], hence the name "systemic IT project management". The following chapter describes the fundamentals of this theory. At first, a short introduction to the variety of systems theories in existence is given, before focusing on the relevant details of Luhmann's theory.

3.1 Fundamentals of Systems Theories

The term "systems theory" can be understood as an umbrella term for theories concerning the description of actions and impacts in systems and society. There are not only one but several different theories and research approaches that can be attributed to a systemic view of the world [7]. Furthermore the research fields of the main representatives of this theory are fundamentally different and are yet complementary in some ways as the subsequent chapters will show. Generally it can be said that the systems theories understand the world as a system consisting of subsystems that are interacting with each other in more or less strong correlation. This relationship is called "structured coupling".

As an example Maturana's and Varela's [15] concept of the "autopoiesis"[1] has been developed in the context of neurobiology. The "Milanese group" [16] specialized in family therapy according to systemic principles. Von Foersters [17] roots belonged to the cybernetics and Niklas Luhmann [5] created a theory about social systems. His theory will be the focus of the following chapters.

3.2 Principles of the Theory of Social Systems by Luhmann

This section will give an overview of the principles of Luhmann's theory of social systems. The first sub-section discusses the different kinds of systems Luhmann differentiates, before the second sub-section looks at the generic concept of "operations" of a system. Of the four kinds of systems social systems are the most relevant for this paper; therefore the third sub-section discusses the key aspect of social systems according to Luhmann: communication. In the final sub-section, the relationship between a system and its environment is examined in the light of his theory.

3.2.1 Types of Systems According to Luhmann

Luhmann [18] identifies four different types of systems: organic systems (e. g. living organisms), psychological systems (e. g. the consciousness), social systems and machines (as long as they are able to observe things). Single humans and groups of humans are – according to Luhmann – no systems by themselves. Moreover, a human being is an aggregation of several different systems. As the title of Luhmann's [18] works already states, social systems are the centre of his theory. These are differentiated in interactions, organisations and societies [20]. Interactions are meant as very short direct contacts only, whereas societies are the largest and most complex persistent social systems which borders are only bound by the restrictions of communication. Organisations are more complex then interactions, are persistent, but can be viewed as subsystems of societies [20].

3.2.2 Systems and Operations

Generally Luhmann states that a system only exists if it is able to operate. Each of the four aforementioned types of systems share the same two basic principles of operation: On the one hand systems just observe. For this reason a machine is a system only if it is able to do so [18], [20]. On the other hand a system differentiates between itself and its environment.

A system according to Luhmann now comes into existence and persistence if operations happen continuously one after another. These operations create and re-create the system itself. This concept is called autopoiesis. A system is autopoietic if its elements are produced and reproduced by the elements it consists of. Every single object used by those systems, like their elements, their processes, their structure and

---

[1] This term will be described in the next chapter.

even themselves are defined through those objects in the system. In other words: There is no input and no output by such objects into and out of the system. This does not mean that there is no relationship to the environment (see section 3.2.5) but those dependencies are located on another level of reality than the autopoiesis itself [5].

3.2.3 Social Systems and Communication

Luhmann now states that the main element a social system consists of and by which it reproduces itself is communication. In other words, communication is the form of operation which constitutes a social system and differentiates it from other types of system [20]. This is an abstraction from humans or their minds – Luhmann's theory states that social systems do not consist of humans or even parts of them, they just consist of communication [20].

Communication always takes place between a sender and a receiver. Every communication process goes through a three point decision process. Three different selections have to be in synthesis to make communication happen as an emergent action. The first selection is about the selection of the information to be communicated, the second is about the selection of the actual message and the third one is the expectation of success [18]. Selection takes place because there are always possible alternatives that can be chosen. A sender first has to choose which information he actually wants to communicate. After this selection he needs to wrap the information into a message. This restricts the information further and requires a media. These two processes are part of the sender. The receiver has the obligation to understand the received message. He has only access to his own perception of the received message and needs to decide if it is relevant for him and if it is, he needs to reconstruct the contained (and intended) information from all possible interpretations. This selection is the most important one because it is the key to a successful communication [18], [20].

The fact that several alternatives exist in each of the three steps makes a successful communication unlikely at first. Luhmann describes such a scenario as follows [18]: Each selection means choosing from a number of different possibilities (selecting what to send, selecting the media, selecting the interpretation of the perceived message). He calls the existence of these possibilities a contingency. Due to the presence of two communication partners with contingencies on both sides Luhmann creates the term "double contingency" [18]. This double contingency however leads to the unexpected effect of lowering the uncertainty of communication because the sender and the receiver get (and need) to observe and to respond to each other in order to ensure and verify a successful communication effort. This creates and re-creates social order [20], contributing to the persistence of the social system.

A further decrease of uncertainty is achieved by three forms of media. The uncertainty in the selection of information is lowered through speech. The uncertainty in communication is lowered through communication media like print, radio, internet or other communication aids. And so called symbolic or success media like money, power, love, law or religion can boost identification of communication and can avoid its refusal. Sense and meaning can be identified as a universal media that applies to all three types of selections [20]. Luhmann differentiates further between three dimensions of sense – the factual dimension, the time dimension and the social dimension [20].

For the self-preservation through continuous re-creation (autopoiesis) of a social system, a continuous flow of communication is necessary. Should there be no more communication happening inside a system its existence is terminated [20].

3.2.4 Systems and Their Environment

Luhmann divides the outside of a system in its direct environment and the other world. He accepts a world that lies outside the reach of a system and disagrees to the position of radical constructivism which negates the existence of such a world. When a system differentiates a part of this outer world then this part turns into a part of the direct environment. This differentiation is an operation by the system. In this perspective, a system is always different to its environment and the environment only exists through the perspective of a system. The environment is the outer face of the system [20].

In order to actually be able to differentiate between themselves and their environment systems need to observe their environment [20]. To perceive and assess differences of any kind an awareness and an understanding has to be achieved, and this can only be developed by observation [21].

Regarding their operations systems are closed to the outside, the exterior to their boundaries. Only an operation of and within a system can generate new operations. Similarly social systems can only generate new communication through previous communication. On the other hand, systems are not closed to external influences. However, the decision if (and how) an external influence will actually affect the system and its operations is always taken by the system itself and cannot be predicted. This openness to the environment allows a structural coupling between systems [20].

4 Principles of Systemic IT Project Management

This chapter now details general principles of systemic project management based on the application of Luhmann's theory on IT projects. At first, three major sources of irrationalities in IT projects are identified and connected to the soft factors mentioned in chapter 2. Afterwards, principles for systemic IT project management are discussed which show possible ways for the IT project manager to deal with these sources of irrationalities. Additionally, necessary changes and consequences for the role and the self-image of the IT project manager are highlighted. And finally, possible barriers for the introduction of systemic IT project management in a project organization are shown.

4.1 Sources of Irrationalities in IT Projects

The application of Luhmann's theory on IT projects as social systems yields three areas which are identified as main sources for irrationalities in IT projects: "diverging perceptions of the truth", "social order" and finally "chaos and order in the project environment".

4.1.1 Diverging Perceptions of the Truth

As already stated in chapter 3.2.5 the systems theory by Luhmann rejects the radical constructivism but the accentuation here is on the word radical. Generally he can be considered as having a constructivist world-view. This means that his systems theory does not believe in only one single reality, but accepts that there are several subjective realities [22]. The key factor here is the observer. Different observers cannot observe the same part of reality in an objective and therefore identical way. Their sense of reality is affected by their point of view, their experiences and their mental constructs. The observers cannot even observe their own application of differentiation criteria during their observation because simply by doing so they already become part of the observed system. For this observation an observer of second order is needed [23].

A closely related theory is the theory of mental models. Mental models are representations of parts of the real world that allow humans imagination, reasoning and logical action. They are also composed of subjective realities [25]. Figure 1 illustrates the basic steps in understanding a sentence via creation and refinement of a mental model.

Words and grammar are the most important aspects regarding the composition of a sentence. In the subsequent steps this information is brought into a context with existing knowledge and experience to create the mental model representing the meaning of the sentence. All this already happens while the sentence is not yet completed but with every single word heard or read [19].

A receiver in a social system utilizes these models for their process of interpreting a communicated message. At the moment the sender initiates a communication effort the receiver uses his existing knowledge in his attempt to understand the meaning and creates his own mental model of it as stated in figure 1. By using this model, further information that was not mentioned before is derived and attached to the model. In order to handle the possible contingencies (see chapter 3.2.3) the receiver needs to

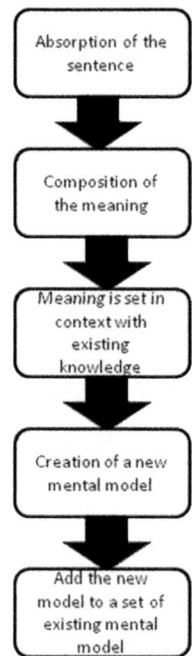

Fig. 1. The main steps in understanding a sentence in natural speech. Source: [19].

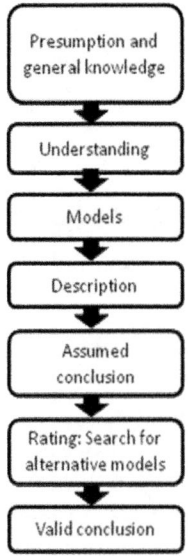

Fig. 2. Three steps of deduction. Source: [24].

create a number of alternative models that can possibly disprove the refined initial model. If it cannot be disproved the model is treated as "correct". Figure 2 visualizes this process completely.

For an IT project this now means that every project member including the project manager will always perceive issues about the project differently than other project members. Furthermore, every project member including the project manager will always be a part of the project system. This prevents him from managing and observing the whole project from the outside and from perceiving a "whole truth".

A popular illustration of this problem in IT project management can be found in figure 3. During the authors' research this was a picture that appeared often and shows the problems of project communication and the different points of view by the example of a swing that shall be attached to a tree. The illustrated different points of view are "How the customer explained it", "How the project leader understood it", "How the analyst designed it", "How the programmer wrote it", "How the business consultant described it", "How the project was documented", "What Operations installed", "How it was supported", and "What the customer really needed".

According to Luhmann's theory this basic problem exists during and after every communication effort inside the social system "project team" and due to the double contingencies on both the sender's and the receiver's side (see chapter 3.2.3) an unambiguous communication effort can virtually not be ensured, regardless whether it is about project goals, requirements (see chapter 2) or other issues. In practice, this can manifest in statements like "But X said that…" or "Who was responsible for communicating Y to Z?"

On the other hand, traditional IT project management methods usually assume and rely on unambiguous communication taking place. This concurs with its mention among the top factors for project success or failure, regardless of the type of project

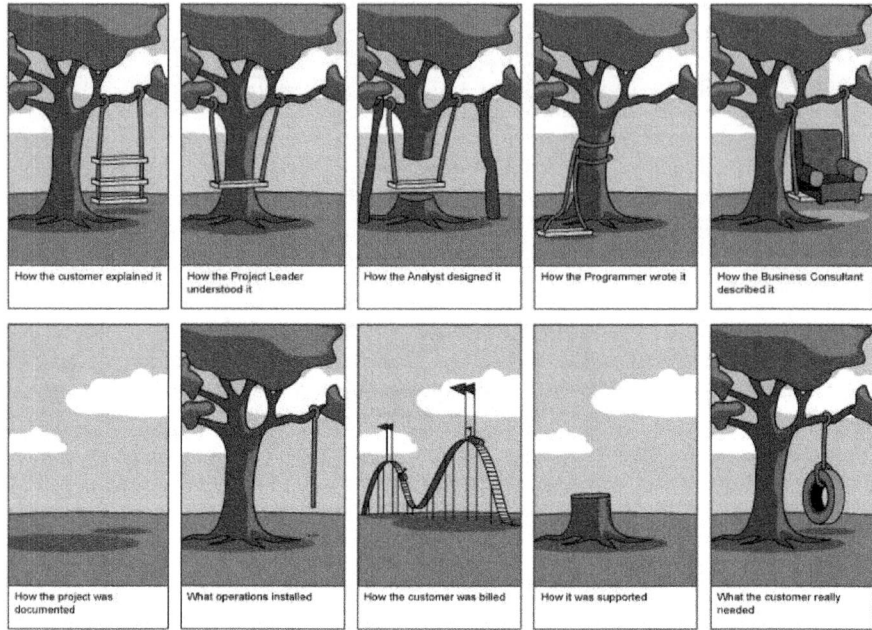

Fig. 3. Project: "Swing". Source: [26], referring to an unknown author.

or method used (see chapter 2.2). Therefore, observed diverging perceptions of a project team member are usually handled as exceptions that need to be "corrected" and therefore might be called irrationalities. Consequences from this first source of irrationalities for project management and the project manager are discussed in chapters 4.2.2 and 4.3.

4.1.2 Aspects of Social Order in Project Management

Chapter 3.2.3 already showed how (abstract) social order is created in social systems according to Luhmann – by continuous mutual observation and reassurance by the sender and the receiver of a communication effort about the communicated information. Among the symbolic media used in these communication efforts in IT project teams are governance, power (by means of the formal project organization), money (regular salary as part of the employment or service contract, incentives) or "truth" (compare the discussion in chapter 4.1.1). These can also be combined; someone responsible for the project budget is able to utilize the media governance and money together. As helpful as an established social order inside the project team may be, the use of these symbolic media may also have negative side-effects, however. The application of formally granted rights or power might be perceived as inappropriate or unfair by team members or stakeholders. Money might be overused as an incentive so that this extrinsic motivation replaces the intrinsic motivation of simply creating the "best possible" IT system. Perceived differences of reality (or "the truth") might not be dealt with sufficiently or at all (see previous section).

These factors related to aspects around social order contribute to the complexity in IT projects. Again, this is an aspect which traditional project management does not consider in an explicit way. Here the focus mostly lies on the "truth" for the reality of the project system as a whole (which doesn't exist, see chapter 4.1.1) and of course the "necessary" formal rights, governance structure and power to lead and control a project team. The negative side-effects of symbolic media like power and governance are neither considered nor compensated in an explicit way. These side-effects of the social order may lead to "politics, department egoisms and conflicts of competences", for example, which chapter 2.2 identified to be among the top three reasons for project failures. Also interpersonal problems, which can be interpreted as disturbances in the continuous mutual reassurance process necessary for successful communication, often lead to negative results [27]. Conflicts between project members are often battled in a hidden way (taking into account the contingencies of communication as means to turn them into a way to deliberately manipulate social order) and finding a consensus, even a compromise, often takes a long time [28].

The authors consider it striking that a good relationship between all project members is not an explicitly identified success factor in the study referred to in chapter 2.2, whereas the negative variant is identified as a key factor for failure. This indicates that this occurrence is also viewed as an irrationality that has to be dealt with outside of the regular IT project management framework. Consequences for the project management and the project manager are discussed in chapters 4.2.3 and 4.3.

4.1.3   Chaos and Order in the IT Project Environment

Today's projects and especially IT projects are characterized by increasing complexity [8], [27]. The methods and tools of classic project management mostly attempt to reduce complexity by imposing order and structure. For example, Gantt charts or similar planning documents attempt to anticipate the length of a project and the resources needed. But despite trying to determine those variables as exact as possible a lot of projects still fail or come into delay.

In Luhmann's terms the social system "project team" tries to observe and differentiate certain parts of its outer world and to turn it into their direct environment (see chapter 3.2.4). The key issue here is that this environment consists not only of the generally observable present but also of the non-observable future (visions, plans etc.). Chaos theory states that a completely deterministic prognosis of proceedings in a process in progress is virtually impossible. In a yet undetermined environment order and chaos depend on each other and their interaction constitutes a dynamic change process. Taken together with Luhmann's theory, there are not only contingencies in the communication within the system, but also a multitude of contingencies in the environment / the outer world of the system. As a consequence, social systems like those in IT projects are regarded as deterministic-chaotic [23].

This chaotic element disallowing an unambiguous observation of the future can also be interpreted as an irrational element since future developments entirely depend on a multitude of contingencies which cannot be anticipated in a comprehensive way. Consequences for the project management and the project manager are discussed in chapters 4.2.1, 4.2.4 and 4.3.

4.2   Principles of Systemic IT Project Management

The following subsections 4.2.2 to 4.2.4 show possible ways of dealing with the three sources of "irrationalities" in IT projects outlined in the previous section. But before this can happen, subsection 4.2.1 elaborates on one more important differentiation - the meaning and the limits of the term "management" in systemic IT project management, especially in the light of the third source called "chaos and order in IT project environments". This is why the subsection headings 4.2.2 to 4.2.4 have the word "managing" set in quotation marks to indicate that it is not the traditional understanding of management which is referred to.

4.2.1   On "Principles" and "Management" in Systemic IT Project Management

Traditionally, the terms "principles" and "management" have a firm and yet safe connotation – principles should allow orientation and adherence with few, if any, exceptions and management is expected to "lead the way" and steer a project (or more generally, the managed organization) to success. In contrast, the deterministic-chaotic nature of a social system "IT project" due to the consequences of "chaos and order in the project environment" (compare chapter 4.1.3) means that any perceived and attributed safety and security of any principles or any managerial effort is an illusion in the end – the uncertainty of the future and the multitude of contingencies mean that theoretically "everything" can happen. Furthermore, since it is up to the social system whether its operations will actually be affected by perceived changes in its environment (chapter 3.2.4), its reaction can be seen as another contingency.

Therefore the perspective on management changes in systemic IT project management – the task is no longer implementing a "plan to success" and avoiding or "fixing" any irrationalities along the way, but leading a project to success through all multitudes of contingencies, uncertainties and irrationalities it may face. The same applies to the principles mentioned in this and the following paragraphs – although they are formulated with these issues in mind, they are certainly not set in stone, and it is the task of the project management to tailor them to the circumstances they perceive in any given moment.

The underlying assumption is, that with acknowledging the existence of the aforementioned sources of irrationalities and also acknowledging the impossibility to make one of the sources stop "generating" irrationalities for the duration of any project, the success chances of any countermeasures – or "interventions" (see next paragraph) – increase. The availability of certain methods for interventions in social systems contributes to this increased success chance. Referring to the title of this paper, the authors see this approach as the more rational way in dealing with project irrationalities than being "surprised" from a project management point of view, whenever irrationalities crop up.

"Countermeasures" in the previous paragraph was another word implying "planned success" – the more suitable term in systemic IT project management would be "intervention". Chapter 3.2.4 stated that the amount of reaction (if at all) of social systems to changes in their environment – even those trying to directly influencing or changing them – cannot be pre-determined. Therefore, a project manager can only try to "intervene" or "irritate" the social sub-systems of their project system to the best of their knowledge and perception, in order to influence their behaviour to be more in accordance to the overall project goals [33]. More about the changing role of the project manager in systemic IT project management can be found in chapter 4.3.

4.2.2 "Managing" Diverging Perceptions of Truth and Reality in IT Projects

After acknowledging the existence of diverging perceptions of truth and reality among the project team members or stakeholders, the question remains how to deal with this issue.

The first step needs to be to uncover the issue in a certain situation. Despite the challenge of the double contingency in every communication effort according to Luhmann (chapter 3.2.3) our own experience with communication seems to indicate that it is usually working "well enough". Diverging perceptions of the truth will probably not be an obvious issue, but the hidden root of conflicts or disagreements about factual project matters at hand.

A first suggestion for dealing with this issue would be to convey the idea of the existence and the possibilities of diverging perceptions to all project team members at the very formation of a project team. This will enable them to apply this principle consciously in their daily business and spot this source of irrationality whenever it might be behind a factual issue at hand. According to the success factors identified in chapter 2.2 this is especially important when dealing with project goals and requirements.

The second step would be to actually resolve the diverging perceptions. This needs support from both the project management (setting aside time and room even within a busy project schedule) and everyone involved (being generally aware of the issue as suggested in the previous paragraph should be very helpful). There are a number of potentially suitable intervention methods available in the literature (for example [38] or [39]). These range from simple question techniques to complex arrangements with organization constellations or role-play. Since it cannot be pre-determined how the social system will react to an intervention (chapter 3.2.4), a more simple method providing less, but a more focused irritation might be more successful as a more complex method. For example, by paying special attention to certain misleading patterns of speech ("The specifications I get *always* lack precision" – "Do you remember one sufficiently precise specification you got? What was it like?"), focusing on possible, desirable future-oriented alternatives ("What would you rather…") or simply turning accusations ("In the past you did…") into personal wishes ("In the future I'd prefer…"), diverging perceptions can be made obvious among the persons involved and dealt with constructively [39].

A pre-selection of intervention methods identified as potentially suitable alongside with training in these methods and an explicit integration in the project management framework used might also be helpful so that in a given situation it is possible to concentrate on resolving the factual issue at hand instead of having to focus on looking for and applying a new and unfamiliar intervention method. An example for an explicit integration of this issue is shown in the MIO framework for systemic IT project management ("lead process") in chapter 5.1.

To be able to execute this process it probably will not be sufficient to rely on monitoring and reporting exceptions "after the fact" – referring to Luhmann's theory a continuous reassurance of unambiguous communication is necessary to prevent the creation of a social order based on unresolved diverging perceptions of the truth which would make a "correction" later on at least difficult.

One barrier to successful application of the suggestions mentioned above is that the project manager and every project member is, by definition, part of the project system. Due to the limits of observation (chapter 3.2.4) a "blind spot" will remain for everyone involved. Since also everyone needs to observe and communicate about possible diverging perceptions, the issue can even occur recursively. In practice, this will probably lead to a deadlock situation during an intervention. In these case, or if the interventions seem to be unsuccessful otherwise, another possibility would be to bring in a true external "observer" observing the social (sub)system in question. This observer of the second order could bring a fresh perspective or can attempt different kinds of interventions (since being a "true outsider") [36], but one should not expect a "flash of undisputable truth" or a sudden "silver bullet solution".

So in the end, this issue has to stay "unsolved" in general, but is usually not "unsolvable" for the persons involved in a specific situation. The responsibility of all people involved to strive to a resolution in specific cases remains. Doing this consciously instead of just intuitively should increase the chances of finding a suitable solution.

4.2.3 "Managing" Social Order in IT Projects

To cope with negative side-effects of symbolic communication media (especially power, governance and money) used and perceived social order in IT projects (see chapter 4.1.2), a dedicated management of consensus, conflicts and cooperation seems to be necessary. Again, one suggestion here is to integrate these steps into the project management methodology like in the MIO approach in chapter 5.1 ("lead process"). The same applies to the recommendation of conscious usage of these media by everyone in the project team with more power or rights than others due to the formal organizational structure of the project. This should include the consideration that the media used is also subject to observation, communication and therefore the possibility of diverging perceptions about their usage exist as well (see chapter 4.2.2).

One aspect we'd like to draw special attention to is the aspect of different types and levels of cooperation in IT projects. In our systemic view, cooperation can be viewed as one manifestation of the existing (or in Luhmann's terms: continuously re-created) social order. Effective, unambiguous communication and mutual trust are essential for an efficient cooperation [29].

In this context, Spieß [29] identifies three different autonomous dimensions of co-operation shown in figure 4.

None of the three dimensions of cooperation are excluding each other. The pseudo cooperation is chosen by Spieß [29] intentionally as the cooperation pictured in the third dimension as explained below.

The background of the strategic form of cooperation is the own benefit of the co-operator from a rational perspective. In cooperation with their partner the co-operator tries to reach a common target that, if each of them would work on their own, would not be reachable, or at least not as easily.

The empathic cooperation has basically the same target as the strategic cooperation. But this time the co-operator tries to put themselves in their partner's position and to what extent they share corresponding interests. The focus here is not so much the target (the corresponding interests) itself, but the process of reaching the common target and the necessary communication involved. To achieve this, perceptions and

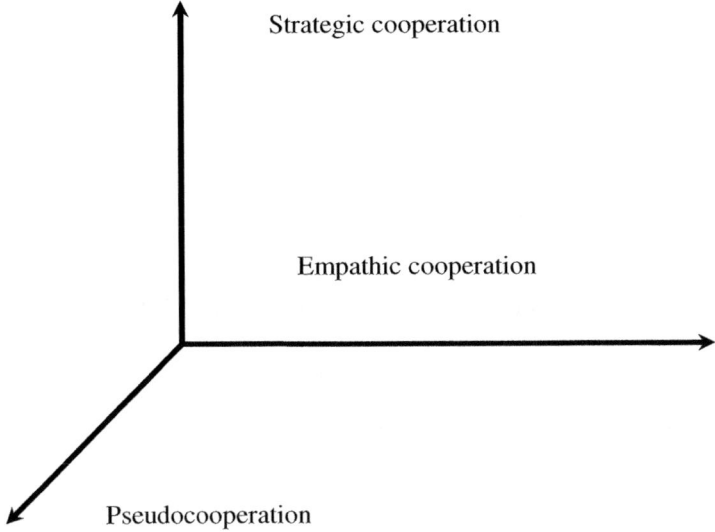

Strategic cooperation

Empathic cooperation

Pseudocooperation

Fig. 4. Forms of cooperation. Source: [29].

feelings, cultural aspects, mindsets and shared practices are relevant. Empathic cooperation is a very important addition to the strategic cooperation since the deliberate focus on the process may help uncover diverging perceptions of the truth. Due to this, it is more communication and interaction intensive [29], which in turn means a potentially more stable social order due to its continuous re-creation.

The transverse pseudo cooperation might emerge during an established strategic or empathic cooperation. The basis for the pseudo cooperation are (self-)deceptions about an existing cooperation which is not effective anymore in terms of mutual benefits. The communication between the partners still occurs but largely only consists of the remains of a previously established social order. In addition, especially the pseudo-emphatic cooperation is said to lead to conditions like the burn-out-syndrome [29] since emotional energy is invested without tangible returns.

A classic example of the pseudo cooperation is work-to-rule where the work is done exactly as prescribed by the current social order but without further initiative or actual benefit for the social system. Another example of pseudo cooperation is the effect of social slacking [29]. In this case, people in a team are working less effectively and with less initiative than the aggregated sum of all individual effort. Additionally, individual responsibilities are shifted to the team so that in the end no one is to blame but the (abstract) team. A cause for social slacking in systemic terms would be the nature of the social order of the social subsystem "team" in combination with the nature of the structural coupling to its superior social system "project" emphasizing the team instead of the team members.

These examples about positive and negative aspects of the different forms of cooperation shall serve to illustrate that the issues related to social order may lie well below the easily observable "surface" of the social system called "project" or its subsystems like "project teams". To determine whether established (and observed) forms of cooperation consist of a useful form or largely of pseudo cooperation, is not an easy one. Since pseudo cooperation usually involves some form of (self-)deception (which might even extend to the most simple forms of pseudo cooperation like work-to-rule, when it happens for a sustained amount of time, for example), only intervention techniques suited for uncovering those deceptions might lead to its detection. Providing a suitable set of methods to detect and deal with pseudo cooperation in IT projects in a systemic way would be subject for an entire paper, however.

But nonetheless, systemic IT project management needs to deal with these issues as well as with the "more obvious" ones mentioned in the first chapter of this subsection, and will serve as a helpful addition to traditional IT project management methods, since it is also in a position to do so in an explicit and well-founded way.

4.2.4 "Managing" Chaos and Order in IT Projects

The third source of irrationalities in IT projects identified in chapter 4.1.3 consisted of the multitude of contingencies how the future could shape the direct environment of a project (which the project system actively observes) or its outer world (which it does not observe). These contingencies form due to the dynamics of chaos and order around the project.

In the light of the existence of these contingencies in our view, it is rational and favourable to acknowledge and accept these dynamics and contingencies instead of trying to reduce the underlying complexity that far so that they can be pressed into a strict plan or concept. Without enough flexibility, the occurrence of unpredictable developments may contribute to project failure in the end [23].

Drawing a link back to chapter 2, unambiguous goals and requirements were mentioned as critical success factors. The aspect of ambiguity can be understood regarding perception (see chapters 4.1.1 and 4.2.2), but also regarding stability over time. Weltz and Ortmann for example regard the gathering of requirement specifications as an active and conscious reduction of uncertainty [28]. That requirements especially in IT projects are "moving targets" which is a well-known issue in requirements engineering (e. g. [40]), but at the same time this means that an effective reduction of uncertainty does not really occur. This leads to the emergence of informal action, contacts and arrangements because they promise an efficient way to a successful project completion, at least in a short-term perspective. Of course, on the other hand this can lead to even more necessary effort due to having to adhere to the formal processes in addition to the informal actions. Another frequently cited issue is a lack of documentation contributing to a failure when even more emergent changes need to be made but prove problematic when the previous emergent changes were implemented informally and without sufficient documentation [28].

From a general management perspective Ortmann [42] advises to make preparations to tackle contingencies directly and approach them as they are. He distinguishes between three types of contingencies. Firstly, a contingency can either concern potential actions (which can be influenced) or occurrences in the environment (which cannot be influenced by the social system in question). Secondly, a contingency can be perceived in a confident (positive, hopeful, constructive, chance-oriented) or a depressive way (as unpleasant necessity or requirement). And finally contingencies differ in the amount of their dependencies of situational factors (something is *contingent upon* something else).

Since a project may face an unpredictable number of contingencies at unpredictable times he recommends enabling the organization to deal with the unplanned and the unexpected – or in his word, to enable their responsiveness. This includes the entire cycle from observation or perception over communication, reflection and finally action within the social system – either as re-action to a contingency which already occurred or as proactive action towards a contingency which is perceived to occur with a certain probability. Depending on the perceived type of contingency a different kind of action is necessary. In the final chapter of his book, however, Ortmann states that "more responsiveness" for an organization is not always better since it could lead to "over-responsiveness" and lack of sense, stability and purpose. In the light of the multitudes of contingencies a project organization may face (chapter 4.1.3), ignoring, forgetting and turning a blind eye is equally important for them in order to retain their ability to exist and act in a meaningful way. He even warns to give one-directional recommendations for organizations to simply become "more responsive".

Here, authors with a strong background in software development projects, like Cockburn [41] provide suggestions which are quite "systemic" in nature and fit well into the general idea of Ortmann's "responsiveness". For example, Cockburn specifically integrates the concepts of requirements "precision" and "stability" in his approach to designing a methodology for a project. Among other things, he advises to strive for only sufficient precision for that another project team can start working with sufficient reliability. Regarding stability, he suggests assessing each team's capability to rework their finished products in case a change crops up later and to design the project specific methodology with this rework capacity in mind. According to him, the project methodology needs to shield teams with the least resources or capacity for rework from change-induced work as much as possible.

So for this aspect, the conclusion is similar to the conclusion of chapter 4.1.2, another "permanently unsolved" aspect, which requires careful, situational solutions from each project team member and additionally, from designers of project methodologies and frameworks, in order not to prevent responsiveness by prescribing a strict process framework. The former aspect requires the creation of a general awareness of the basic issue among the members of the project team who have to be up to the challenge of dealing with it throughout the lifetime of the project.

4.3 Redefinition of the Role of the IT Project Manager in Systemically Managed IT Projects

After looking at the direct consequences of the sources of irrationalities for the social system called "project", this chapter takes an institutional perspective and looks at new requirements or recommendations for responsible IT project manager(s).

Unless a systemic approach to IT project management is already prescribed by the corporate project management framework, it is the personal responsibility of the project manager to take the first steps to implement and try systemic IT project management in addition to the "regular" project management. Furthermore, it is his responsibility to decide, to what extent the principles and recommendations of systemic IT project management should be followed and how the previously recommended familiarization of the team members with the systemic principles should take place.

To ensure a strong anchor against the currents of the hectic daily business, the recommendation from previous chapters is repeated here, to integrate the systemic principles into the used project management methodology to make the new approach visible to every project team member.

The direct consequence of the first source identified in this paper for the project manager is that he is not always right and that there is a solid chance that one or more project team members perceive them with good reasons as being wrong. This leads to the requirement of knowing when to allow (or actively seek and encourage) opposition and constructive discussion to ensure both unambiguous perceptions and unambiguous and shared ideas and goals.

The direct consequence of the second identified source is that the project manager needs to be aware that while communication creates and re-creates social order, their choice of symbolic media can heavily influence this social order. The permanent re-creation of social order through communication also means that it is not carved in stone but has the potential of changing continuously. Additionally, this social order (or in other words, the informal organization) needs at least as much personal attention and care as the formal project organization.

The direct consequence of the third identified source finally is the insight, that the project manager is limited in his actions towards the project system. Both the existence of a multitude of contingencies as well as the unpredictable dynamic of any social system regarding their reaction to interventions or irritations from their environment prevent a mechanistic or cause-and-effect oriented approach of being successful in the long run. In other words, a project manager must not ever fall victim to the illusion of being in complete control. However, this is not a recommendation to react in an entirely depressive way against every contingency (see chapter 4.2.3) and to forgo any leadership as a consequence. The key here is to strike a balance between showing personal uncertainty and, on the other hand, actively following up on actionable contingencies (opportunities) in a confident way or at least consciously providing the illusion of a clear direction. Kühl calls this shifted purpose of management "deliberately putting the own organization's mind at ease" [37]. The important difference to the traditional way of management is not only knowing not to be in complete control all the time, but also knowing when to pretend to be, in order to achieve a hopefully positive effect on the social system "project".

In conclusion, this means, the basic understanding of the role of project manager needs to change as well in accordance to the aforementioned principles (and the new understanding of the terms "management" and "principles" outlined in chapter 4.2.1) - from authority figure to coach, source of inspiration or development worker, depending on the situation ([34], [35]). A magazine incorporated this "paradigm shift" even into its title "Review of Post-heroic Management" (in German: "Revue für post-heroisches Management") – associating the traditional managerial role with the role of the classic "hero" while the changed managerial role is dubbed "post-heroic".

4.4   Barriers to Systemic Project Management in Practice

After the discussion of principles of systemic IT project management and the changed role of the project manager this final section of the fourth chapter will take a look at possible barriers to systemic IT project management in practice.

One barrier could be the necessity of integration into existing project management frameworks, especially if those are mandatory to use and/or lack a certain flexibility or responsiveness to integrate the additional systemic elements. An example of an extended project management framework can be found in chapter 5.1.

A second barrier to overcome consists of personal habits of every member of the project team as well as influences from the industry, corporate, team and/or professional culture. The basic systemic approach is not necessarily intuitive for someone not familiar with the underlying theories, and some elements of certain cultures may be a hindrance or lead to an outright refusal of ideas like differing perceptions, unpredictable contingencies or the importance of the social order.

Another criticism towards Luhmann's theory in general and approaches based on it is a certain vagueness regarding terms and definitions as well as the necessity to apply and tailor the general principles to the specific circumstances. But on the other hand exactly this is an intended strength of the systemic approach, not to promise a "one-size-fits-it-all" approach, but to provide the necessary adaptability to fit into a specific project environment and to empower the people involved to conduct the adaptation.

To support the argument for a viability of a systemic approach to IT project management a comparison to traditional methods of project management without systemic elements would be necessary. Due to the uniqueness of each project situation, a strict comparison is virtually impossible. Even formulating key performance indicators (KPIs) to measure isolated success of the systemic elements is a challenge at best. Due to the unpredictability of contingencies (see chapters 4.1.3 and 4.2.4) and the relevance of subjective perceptions and observations (see chapters 4.1.1 and 4.2.3) finding scales or criteria for constructing such KPIs is certainly not a trivial task. Only a certain feasibility could be inferred from trends of increasing project success rates or the subjective perception of project team members and managers whether the additional effort and methods of systemic project management were suitable and helped a project to its success.

And finally, systemic approaches in IT project management are yet mostly untried in practice, although some authors especially from the area of agile project management like Cockburn (see chapter 4.2.4) "intuitively" tend to follow some systemic principles when giving recommendations or designing project methodologies.

5   Variants of Systemic Project Management in Practice

The following three systemic project management approaches can be found in the literature and are presented and discussed in order to show a variety of ways how systemic IT project management can be implemented in practice instead of creating "yet another approach".

5.1   The MIO Approach of the University of Zurich

The development of the MIO approach started in 1997 at the institute of informatics at the University of Zurich. Since then it was consistently further developed further and even evolved to the study degree: "Diploma of advanced studies in IT project management". "MIO" is the abbreviation for the German words "Mensch – Informatik – Organisation" (human – informatics – organization) [8], [9], [10].

The approach takes the high dynamic and complexity of IT projects into account. Major changes in the project environment or the project team in consequence of internal as well as external factors are addressed explicitly [8]. Therefore it accepts the lack of predictability and the existence of contingencies in IT projects. It also incorporates elements of chaos theory [45]. Project complexity is explained through the existence of different project stakeholders, their opinions, visions and influences. This constellation of authority and demand might result in conflicts of interests which result from social problems and personal aversions [8], [45].

This shows that the already discussed social aspects can be found in the MIO concept. The diverging perceptions of truth are not included in an explicit manner, but can be found implicitly. Huber and Kuhnt are talking about the term "observation" as encoded, individual, cognitive mindmaps which are passed along in form of numbers, speech, texts and pictures [9]. This term correlates with the "mental models" discussed in chapter 4.1.1.

To deal with this situation the MIO approach visualizes IT projects as composition of three parallel processes. There is a project management process which addresses the classic concepts of project management. Starting with a project charter a project gets initialized and planned, the realization takes place including a controlling function and it ends with the project close-out. The project charter is also trigger for the second process, the product development process. Known examples for this process are the phase models like the waterfall model, the spiral model or the model V XT of the German government.

The new and additional concept in the MIO framework is a so called lead process. This lead process focuses on social and soft factors, or in other words, the irrationalities, in a managed project. It is divided into four phases: An establishment-phase to analyze possible problems in the project team and the environment. A constitution phase, including a "kick-off workshop" for example, to build the team and lay a foundation for unambiguous communication. A conduction phase is next, in which the project lead has to fulfill tasks of motivation, mediation, marketing to the stakeholders and course corrections if something goes not the desired way. These tasks need to take place within the project and the surrounding environment. In case of rapid changes in the project or its environment certain "serenity" is advised regarding the chaotic-deterministic elements of IT projects. It is suggested to reflect the project after its ending, for example in a "kick-out workshop" to learn from it. The following figure shows the three processes and their relations to each other.

Besides those three processes Kuhnt [11] considers IT projects in its systemic context. A separation between the project system and the customer system for example through an organizational or local split might lead to communication difficulties. The two systems would only be able to intervene in their respective counterparts but not affect themselves directly. For this reason the executive persons of the project system should be embedded in the customer system.

Furthermore the IT project should be counseled by a social advisor. This advisor has to guide the team in non-technical topics of the mentioned lead process. These topics might be reducing communication barriers, representation of the project to the external stakeholders and early diagnostics of possible problems [9], [11]. With the self-perception of a "chief servant" instead of a domain expert the advisor helps the project management to anticipate chaotic times during the assignment and avoid conflicts [11], [8].

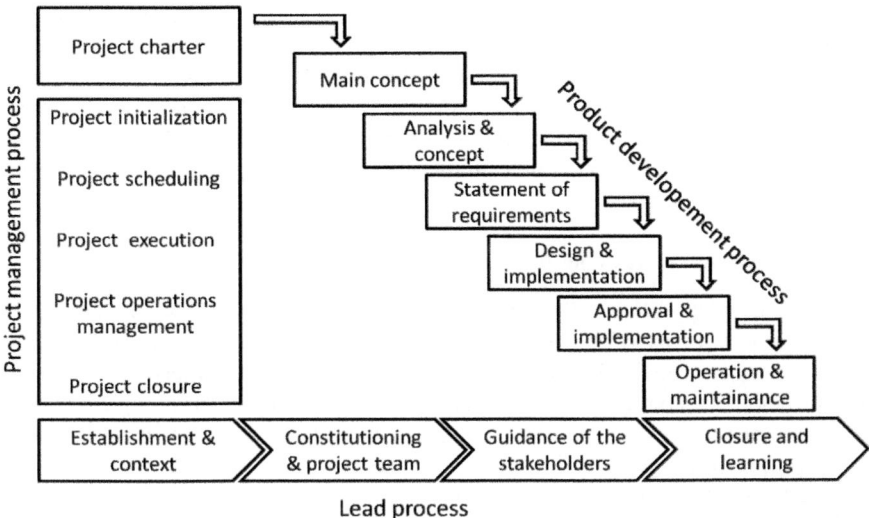

Fig. 5. The three processes of IT projects according to MIO. Source: [9].

Additionally to these guidelines the MIO-approach describes six social success factors for a satisfying IT project management and controlling. These factors are:

- Creation of a project identity with its own values, social norms, visions and rituals to clarify the reason of the project to the project team.
- Observation of the handling of information by the project system, so an early intervention is possible in case of communication difficulties.
- It is important to learn and refine the storage and reuse of information throughout the project.
- Knowledge of the balance of power to make it applicable.
- The factor time is a complex point in IT projects. The MIO approach considers it on three levels: There is time seen as resource, which can be manipulated through manpower, for example computing time, financial resources or other concepts. Then there is time as a fixed point (for example a milestone deadline) which cannot be manipulated. The third level is time as system time which cannot be affected directly since a project system needs it's time to fulfill certain kinds of processes, but can be supported by good framework conditions. This corresponds to the first and third kind of contingencies by Ortmann mentioned in chapter 4.2.4.
- IT projects are usually interdisciplinary projects. For this reason there must be a focus on the collaboration of different people and their qualifications [45].

Although this approach delivers several interesting aspects for managing IT projects in a systemic way and incorporates many of the factors discussed in this paper, there is still room for improvement. Looking at the three processes that take place in an IT project only the elements of the lead process are used to implement change in a systemic way. It should be questioned if an adaption of the other processes would make sense as well or is even necessary. The question is whether it is feasible to "outsource" all social aspects into the lead process and keep the other two process realms as they are. Especially the change to a human centered project management could probably be a good addition to cope with uncertainties and irrationalities. Besides that the MIO approach addresses the product development process in form of the more classical phase models like waterfall model or spiral model. It does not consider modern agile software development methodologies like Crystal [41], SCRUM [43] or Extreme Programming [44] which might already include parts of the lead process as hinted in chapters 4.2.4.

5.2 The St. Gallen Approach

The St. Gallen approach following is based on a paper by Schwaninger and Körner [12] of the University of St. Gallen published in 2001. The full title is "Systemisches Projektmanagement: Ein Instrumentarium für komplexe Veränderungs- und Entwicklungsprojekte" (systemic project management: an instrument for complex change- and development projects) and unlike the MIO approach it is not explicitly an approach for IT project management. Instead, the focus lies more on a general technical and engineering-oriented approach, which is consistent with typical IT projects, however. And as well as MIO the St. Gallen approach takes the position that the classic project management cannot handle complex IT projects all on its own.

The approach criticizes the typical paradigm of deterministic causal reasoning. This concept is common in the western hemisphere, especially in the area of management and tries to solve every behavior or occurrence using a direct relation between the cause and consequence and does not consider the systemic surroundings [12]. The St. Gallen approach also states that projects and their environments are dynamic and the planning must not be inflexible but fast adaptive to new circumstances [12]. These two viewpoints correspond to the principles of systemic IT project management outlined in chapter 4 of this paper.

In addition to that the St. Gallen approach identifies social factors as a possible cause of problems and predicts that not individual improvements can boost the project but the relations of the team and stakeholders among themselves. Therefore communication is important [12]. Besides that the approach questions the feasibility of hierarchic project organizations since they slow down the flow of information [12]. Finally a so called normative management should support the project to ensure the right motivation and mentality [12].

The St. Gallen approach can be divided in six procedures while one of these is not considered systemic and will not be discussed further. At first, the primary processes of a project are seen in a systemic view of circulating processes as opposed to having a relationship of cause and effect. These processes can be influenced positively through interventions. To be able to do that the authors of the St. Gallen approach recommend visualizing the core processes of the project. The interventions should aim on long-run positive economic and social changes, but the authors highlight that success cannot be guaranteed. This corresponds to the behavior of social systems described in chapter 3.2.4 in this paper.

According to the St. Gallen approach the best performance might be achieved when the project system is viewed a holistic way. For example, there is a high probability that there is a close relationship between the provision of the financial resources and the amount of political influence. Through this structural linkage both concepts can influence themselves in a positive or in a negative way [12].

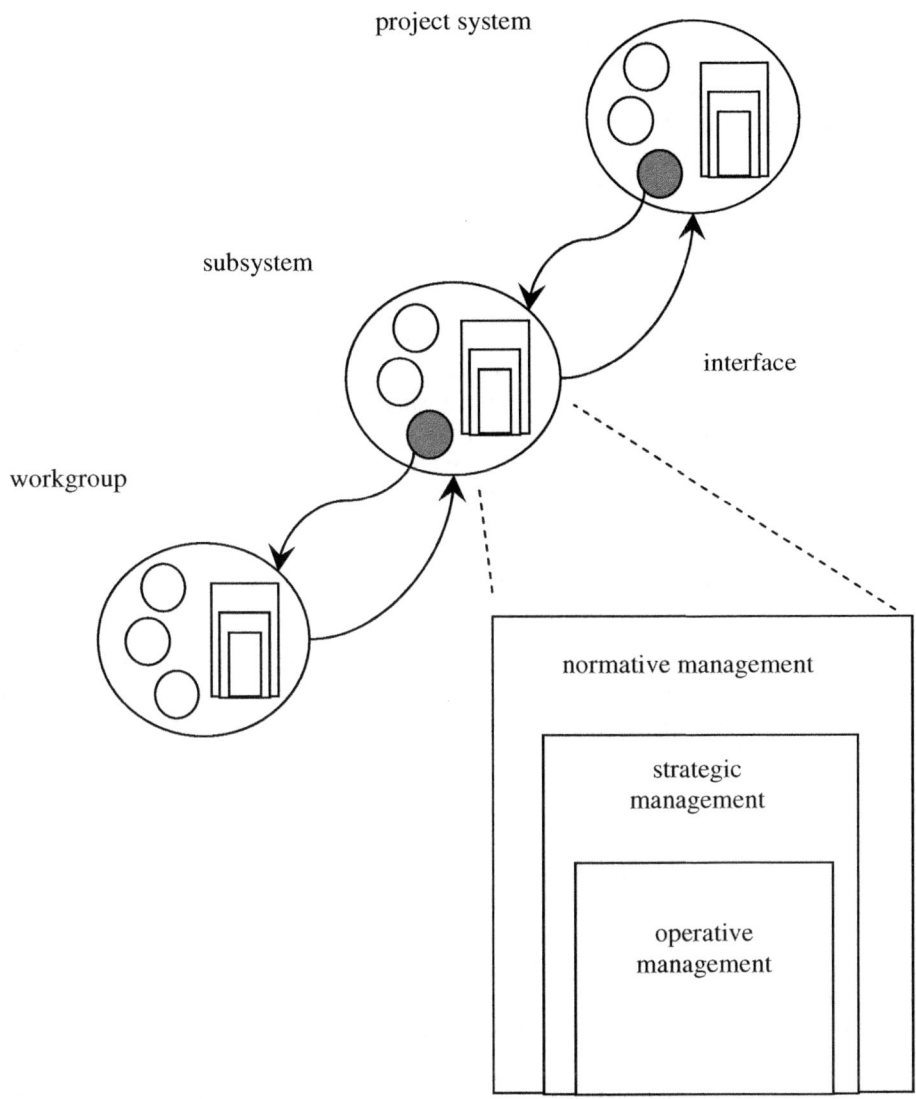

Fig. 6. Model of the recursive management including the three-tier-management-model. Source: [12].

Schwaninger und Körner approach the topic of project complexity in a very thorough way. Two procedures aim to reduce complexity. Those two are the decomposition of the project systems in subsystems and the installation of a normative management in each system. The following figure shows the fragmentation of the complex project system in its parts and the connection of them via interfaces. This way the authority and responsibility is delegated and the motivation of the team members is likely to rise. Also in each system a normative management should be installed besides the strategic and operative management. This management is supposed to ensure the long term success and watch the conditions of the cultural framework [12].

The fourth procedure considers the project dynamics. The relation between operative, strategic and normative management as well as the surrounding environment needs to be determined. These relations are essential for the project but they are also very dynamic. Therefore the planning of the project has to be flexible as well, has to consider instabilities and changes and therefore needs to be adapted continuously [12].

The sixth procedure deals with possible consultants to the project team. The project system and the consulting system can be seen as a combined system itself. The consulting part has to be as close to the project system to be able to intervene if necessary but sufficiently far away to retain a neutral stance [12].

This approach makes a couple of recommendations for systemic IT project management like a flexible and dynamic project planning, the concentration of improved relations among the project team and the stakeholders rather than individual improvements as well as establishing a normative management. All these points seem clear and correct but it is not said how to integrate them into a systemic project management. The authors are also missing detailed instructions how to follow these guidelines in practice.

5.3   Systemic Project Management According to Sumetzberger

As already stated in chapter 4.4 practical experiences with systemic approaches to IT project management are very limited at the moment and scientific studies about the viability of such approaches remain one large "blind spot" for the research community. For this reason and as opposed to the two previous approaches with a strong scientific background, this chapter will show a practical approach, based on a conference paper by Walter Sumetzberger [13] from the systemic consultancy firm called OSB-International. This approach covers the self-perception of projects, their communication processes and their dimensions of intervention.

The self-perception in a project is relevant for the information flow and coordination. Projects can only be controlled through the awareness of ones own steering actions and the knowledge about possible failure. For this reason it is necessary to obtain all relevant information for the self-perception and to analyse them. The information needed is depending on the goals of the project. These goals are not just the classical triangle goals cost, time and quality but also soft factors like organizational dynamics, communication processes, qualifications and visions. Many tools and instruments of the classic project management allow a good view on the past but this is not the essential perspective in a project, since it is usually a unique event. As also concluded in chapter 4.2.4 of this paper, anticipation is necessary. It is relevant to know if the project is running well at the moment and in the future, or if interventions might be necessary. Therefore the right information should be inquired at the right point and the right time. These elicitations must stay flexible to give room for creative influences while steering the project. The key is staying as dynamic as the project will be [13].

This information gathered should be communicated to the people who need them. How to do this is very important since even "taboo topics" must be addressed, but in a sensitive manner. For this purpose a good insight into human nature, empathy and the ability of being a good listener is helpful. Additionally the differentiation of important and unimportant information for the audience is crucial [13].

Referring to Simons [14] Sumetzberger suggests four dimensions of steering interventions, which should influence the project in an indirect manner.

The first dimension is a cognitive system of beliefs which is a combination of values, mindsets and visions of the project. The management is supposed to lead the project team by example regarding these beliefs. The team now can gain orientation observing and following this example, in case of questions regarding values etc. The second dimension includes rules for the project system. Already during the formation of these rules and their possible sanctions an intervention takes place. The third dimension is a system to determine deviations from previously set targets during the project. The intervention in this dimension takes place already during the specification of the relevant observation points. Also the adjustment actions in case of deviations are considered interventions. The last dimension is a steering system, which should monitor important strategic project factors. These factors, especially the problematic ones, should be communicated throughout the project team early to prevent failures [13].

Although Sumetzberger introduces a whole new approach for the leading and steering IT projects he does not give any practical advice how to introduce such a new way to an existing management infrastructure. In his explanation he states that the classic way of steering just looks at the past and that in a systemic way the present is in focus. Nonetheless it seems useful to implement this new approach in parallel to the old way of steering where historic data determine how things should turn out in the future. Maybe using both perspectives together with all relevant data would be a more effective approach than using just either perspective.

6 Systemic IT Project Management and Enterprise Transformation

This chapter will, albeit briefly, apply the principles discussed so far to the fields of organizational change and enterprise transformation. The first section will take a generic perspective, while the second section will take an exemplary focus on the integration with an existing approach of enterprise transformation.

6.1 Generic Application of Systemic Principles to Enterprise Transformation

Luhmann's theory is considered a universal theory for social systems [20]. This allows an easy transfer of the aforementioned principles for systemic IT project management to the topic of enterprise transformation or organizational change. For a project manager of a project organization this becomes immediately relevant when he wants to introduce the principles of systemic IT project management and therefore change established ways of project work.

The new social system in question would now be the organization or the part of the enterprise subject to transformation. The departments, divisions etc. within that part or enterprise are considered social subsystems, just like project teams inside a project organization.

Now our findings can be transformed as follows. Diverging perceptions of the truth also exist inside each social (sub)systems, but this time not only of the "truth in the present" (current organizational reality) but also about the "truth" of the intended changes or the transformation goals. Subsequently, these diverging truths have to be taken care of by change management. Secondly, as social order is about to be changed, those who perceive a potential loss from the perceived new social order will initiate communication efforts (verbally or by means of other symbolic media at their disposal, like power or money) to keep the previously established social order. Thirdly, the initiative of a transformation of an enterprise can be considered an intervention which results are unpredictable in the end. And finally, the changing role of the project manager transforms into a changed role and self-understanding of the person(s) leading and governing the transformation process, turning similarly from authority figure(s) into coach, source of inspiration or development worker (see chapter 4.3).

6.2 Exemplary Application of the Systemic Principles to an Existing Enterprise Transformation Approach

This section will provide an exemplary first link between the systemic approach presented in this paper and an enterprise transformation approach from the literature, in order to show a possible integration beyond the generic application of the principles in the previous subsection. The chosen approach is called "Informed Governance of Enterprise transformations" by Harmsen, Proper and Kok [46].

They view the process of enterprise transformation on three levels – the "operational enterprise system" (basically the enterprise which should be transformed), the "enterprise transformation system" (which is "in charge" of governing the actual transformation processes) and the "transformation maturation system" (which aims to continuously improve the enterprise transformation system and guide it to a higher level of maturity). The actual "enterprise transformation" effort is depicted as a single arrow leading from the "enterprise transformation system" to the "operational enterprise system". There is a similar arrow leading from the "transformation maturation system" to the "enterprise transformation system", indicating another transformation effort, this time directed at the latter.

Assuming the viewpoint of this paper here, the "operational enterprise system" is the equivalent of what we would call the social system "enterprise". An application of the systemic principles to it was already discussed briefly in chapter 6.1. The "enterprise transformation system" basically is the equivalent of the project team in charge of the transformation processes, so all principles outlined in this paper would directly apply to this social system "project team" as well. The "transformation maturation system" has no direct, structural equivalent in this paper here. Only the necessity to consistently change and adapt in order to handle "chaos and order" in the environment was mentioned. It is certainly a prudent measure to try to do so in an explicit and controlled way – to intervene on one's own social system, in a way. And finally, a systemic perspective on the transformation efforts would regard them as a multitude of interventions, highlighting the unpredictable nature of external influences and the lack of guaranteed success towards the target systems in question.

There were several other aspects mentioned in [46] which could be directly linked to the systemic principles outlined here (e. g. communication styles or attitude of the enterprise architect); a more detailed discussion is unfortunately beyond the scope of this paper. Nonetheless, the previous paragraph showed basic starting points for an integrated perspective through a combination of the two approaches. While the approach by Harmsen et al. focuses on transforming an enterprise in a controlled and governed way, the systemic principles focus on the irrational or "uncontrollable" aspects of such processes. The authors think that further research could well show complementary (and more concrete) ways to integrate the two basic "philosophies" of viewing enterprise transformation.

7   Conclusions and Outlook

Based on critical factors for success and failure in IT projects and Luhmann's theory of social systems this paper showed three sources of irrationalities in IT projects as a foundation for principles of systemic IT project management in order to deal with these irrationalities. To provide further illustration and applicability three systemic frameworks for IT projects were also presented and discussed.

The three sources of irrationalities are comprised of "diverging perceptions of the truth", "the social order in project teams" and finally "chaos and order in the project environment". Consequences for the project management include uncovering and resolving the different perceptions of the truth, special attention to social order related issues like cooperation and the necessarily of responding to a multitude of contingencies through organizational responsiveness. The project manager also faces a role change, from the person in full control of the project, authority figure and "hero" to a "post-heroic" advisor, coach and servant to the project as a whole. Furthermore, the social system "project" can only be influenced through irritations and interventions whose effectiveness cannot be predicted. And finally, project frameworks like the MIO framework in chapter 5.1 showed the possibility of extending traditional methods and frameworks for IT project management with systemic elements. Additionally, it was shown in an exemplary way how the systemic principles could be applied to the field of enterprise transformation itself.

Desiderata for future research consists in a further criticism and refinement of the principles of systemic IT project management presented in this paper, the design and evaluation of IT project specific systemic methods and interventions as well as the evaluation of applications and cases of systemic IT project management throughout entire projects in practice. The same applies for enterprise transformation approaches, methods, and projects, as outlined in chapter 6.

References

1. Standish Group: CHAOS Summary Report (2009)
2. Standish Group: The CHAOS Report (1994)
3. Krcmar, H.: Informationsmanagement. Springer, Heidelberg (2010)
4. Sommerville, I.: Software Engineering. Pearson, Munich (2007)
5. Luhmann, N.: Soziale Systeme. Suhrkamp, Frankfurt a. M. (2008)
6. CIO.com: Recession Causes Rising IT Project Failure Rates, http://www.cio.com/article/495306/Recession_Causes_Rising_IT_Project_Failure_Rates (accessed 2009/09/02)

7. Luhmann, N.: Einführung in die Systemtheorie. Carl-Auer, Heidelberg (2002)
8. Huber, A., Kuhnt, B.: IT-Projektmanagement – Kommunikation als Schlüssel zum Erfolg. Informatik-Spektrum 28(4) (2005)
9. Huber, A., Kuhnt, B.: IT-Projektmanagement – Eine kommunikative Herausforderung. FIfF-Kommunikation, 1/07 (2007)
10. Institut für Informatik, Universität Zürich: MIO-Web, http://www.ifi.uzh.ch/arvo/mio/diplomkurs/dsg0607.html (accessed on 2010/05/24)
11. Kuhnt, B.: Systemische Beratung in kooperativen Softwareprojekten. Informatik Spektrum 20(1) (1997)
12. Schwaninger, M., Körner, M.: Systemisches Projektmanagement: Ein Instrumentarium für komplexe Veränderungs- und Entwicklungsprojekte, Diskussionspapier, vol. 43. University of St. Gallen (2001)
13. Sumetzberger, W.: Neue Herausforderungen für das Projektcontrolling aus systemischer Sicht. In: InterPM Conference Glashütten (2004)
14. Simons, R.: Control in an Age of Empowerment. Harvard Business Review (2) (March/April 1995)
15. Maturana, H., Varela, F.: Der Baum der Erkenntnis. Fischer, Frankfurt a. M. (2009)
16. Palazzoli, M., Selvini, L., Boscolo, G., Cecchin, Prata, G.: Paradoxon und Gegenparadoxon - Ein neues Therapiemodell für die Familie mit schizophrener Störung. Klett-Cotta, Stuttgart (1981)
17. von Foerster, H.: Das Konstruieren einer Wirklichkeit. In: Watzlawick, Paul (eds.) Die erfundene Wirklichkeit, pp. 39–60. Piper, München (1998)
18. Luhmann, N.: Soziale Systeme. Grundriß einer allgemeinen Theorie. Suhrkamp, Frankfurt a. M. (1987)
19. Sanford, A.: The nature and limits of human understanding. T&T Clark, London (2003)
20. Berghaus, M.: Luhmann leicht gemacht. Böhlau, Köln (2004)
21. Kuhnt, B.: Softwareentwicklung als systemische Intervention in Organisationen. Ph.D. thesis, University of Zürich (1998)
22. Ritter, J., Gründer, K.: Historisches Wörterbuch der Philosophie, Band 10: St-T, Schwabe & Co. AG, Basel (1998)
23. Ellebracht, H., Lenz, G., Osterhold, G., Schäfer, H.: Systemische Organisations- und Unternehmensberatung. Praxishandbuch für Berater und Führungskräfte. Gabler, Heidelberg (2003)
24. Johnson-Laird, P.N., Byrne, R.M.J.: Deduction. Lawrence Erlbaum Associates Ltd., Mahwah (1991)
25. Johnson-Laird, P.N.: How we reason. Oxford University Press, Oxford (2006)
26. ITligenz: ITligenz: Projekt "Schaukel" (2005), http://itligenz.twoday.net/stories/1223279/ (published on 2005/12/02) (accessed on 2009/11/21)
27. Gudacker, M., Eisenblatt, IT-Management (Teil I): Systemisches Projektmanagement: Der Mensch macht's. Vorstellung eines Instruments zur besseren Bewältigung der sozialen Komplexität und Dynamik von IT-Projekten. In: Ordix News. Das IT-Magazin der Ordix AG, 02/2009, Paderborn, pp. 38–41 (2009)
28. Weltz, F., Ortmann, R.G.: Das Softwareprojekt – Projektmanagement in der Praxis. Campus, Frankfurt (1992)
29. Spieß, E.: Kooperatives Handeln in Organisationen. Theoriestränge und empirische Studien. Rainer Hampp, Mering (1996)
30. Specker, A.: Modellierung von Informationssystemen: Ein methodischer Leitfaden zur Projektabwicklung. Vdf Verlag, Zürich (2005)

31. GPM e. V.: Ergebnisse der Projektmanagement-Studie 2008 – Erfolg und Scheitern im Projektmanagement, GPM e. V.,
    `http://www.gpm-ipma.de/fileadmin/user_upload/`
    `Know-How/Ergebnisse_Erfolg_und_Scheitern-Studie_2008.pdf`
    (accessed on 2009/11/17)
32. Kotulla, A.: Management von Softwareprojekten: Erfolgs- und Misserfolgsfaktoren bei international verteilter Entwicklung. Gabler, Wiesbaden (2002)
33. Kasper, H., Mayrhofer, W., Meyer, M.: Managerhandeln – nach der systemtheoretisch-konstruktivistischen Wende, Die Betriebswirtschaft, pp. 603–621 (May 1998)
34. Baumgartner, P., Payr, S.: Lernen mit Software. Österreichischer Studienverlag, Innsbruck (1994)
35. Königswieser, R., Hillebrand, M.: Einführung in die systemische Organisationsberatung. Carl-Auer, Heidelberg (2007)
36. Backhausen, W.: Management 2. Ordnung. Chancen und Risiken des notwendigen Wandels. Gabler, Wiesbaden (2009)
37. Kühl, S.: Das Regenmacher-Phänomen: Widersprüche und Aberglaube im Konzept der lernenden Organisation. Campus, Frankfurt (2000)
38. Königswieser, R., Exner, A.: Systemische Intervention. Architekturen und Designs für Berater und Veränderungsmanager, Klett-Cotta (2008)
39. Prior, M.: MiniMax-Interventionen: 15 minimale Interventionen mit maximaler Wirkung. Carl Auer, Heidelberg (2000)
40. Macaulay, L., Mylopoulos: Requirements Engineering: An Educational Dilemma. Automated Software Engineering 2, 343–351 (1995)
41. Cockburn, A.: Agile Software Development. Pearson, Boston (2002)
42. Ortmann, G.: Management in der Hypermoderne – Kontingenz und Entscheidung. VS Verlag, Wiesbaden (2009)
43. Schwaber, K.: Agile Project Management with Scrum. Microsoft Press, Redmond (2004)
44. Beck, K., Andres, C.: Extreme Programming Explained – Embrace Change. Pearson, Upper Saddle River (2006)
45. Huber, A., Kuhnt, B.: IT-Projekte als soziale Systeme. In: Setzwein, C., Setzwein, M. (eds.) Turnaround-Management von IT-Projekten. dpunkt, Heidelberg (2008)
46. Harmsen, F., Proper, H.A.E., Kok, N.: Informed Governance of Enterprise Transformations. In: Practice-driven Research on Enterprise Transformation, pp. 155–180 (2009)
47. Bon, J.v., Verheijen, T.: Frameworks for IT Management. van Haren, Zaltbommel (2006)

On Supporting Collaborative Problem Solving in Enterprise Architecture Creation

Agnes Nakakawa[1], Patrick van Bommel[1], and H.A. Erik Proper[1,2]

[1] Institute of Computing and Information Sciences, Radboud University Nijmegen
P.O. Box 9010, 6500 GL Nijmegen, The Netherlands
[2] CITI, CRP Henri Tudor Luxembourg, Luxembourg
A.Nakakawa@science.ru.nl, pvb@cs.ru.nl, e.proper@acm.org

Abstract. Creating enterprise architecture can be perceived as a creative problem solving task, since it involves managing organizational complexity and inflexibility by devising a synergic solution from all organizational units. Creative (or collaborative) problem solving in several fields has been supported by supplementing domain specific techniques with functionalities of a Group Support System (GSS). This paper aims to demonstrate how GSSs can also be used to support collaborative problem solving in enterprise architecture creation. Using the Design Science research methodology, a method was designed to support collaborative problem solving during architecture creation. This method draws from enterprise architecture approaches that are used in practice, and collaborative problem solving theories in academia. It has been evaluated using an experiment and two real life cases. This paper presents findings from this evaluation. The findings were used to refine the method, and they indicate that the effectiveness of academia-based artifacts in addressing problems encountered in practice, can only be achieved through continuous and diverse evaluation of these artifacts in practice.

Keywords: Enterprise Architecture Creation, Collaborative Problem Solving.

1 Introduction

Alignment between an organization's business and IT strategies enables it to realize value (or improved business performance) from its IT investments [8]. However, this alignment is not enough, as there is need to align human, organizational, informational, and technological aspects of an organization [24]. Aligning all these aspects requires using enterprise architecture [29,24], or an integrated or multi perspective approach [15,36]. With enterprise architecture, an organization is able to manage the complexity and inflexibility of its business processes, information systems, and technology infrastructure [27]. Enterprise architecture addresses enterprise-wide integration [15]. Thus, creating enterprise architecture requires formulating a synergic solution from all organizational units. This synergy of the various capabilities in an organization enables it to acquire a sustainable competitive advantage [31].

F. Harmsen et al. (Eds.): PRET 2010, LNBIP 69, pp. 156–181, 2010.

Creating enterprise architecture generally involves: creating a joint conceptualization of problems, strategies or solutions [24]; identifying and refining stakeholders' concerns and requirements; developing architecture views that show how these requirements will be addressed, trade-offs that need to be made to resolve any conflicts [35]; assessing alternatives; risk assessment and mitigation; making decisions [24]; and communicating the architecture [26,24]. On the other hand, collaborative problem solving (or decision making) involves: having direct and reciprocal communication (about the situation at hand) among parties involved; being creative in formulating solution strategies and new alternatives; making shared decisions; and reaping joint payoffs from the decisions made [28]. It can be noted that the enterprise architecture creation activities mentioned above, involve collaborative problem solving activities. Thus, enterprise architecture creation can be perceived as a collaborative (or creative) problem solving task. Collaboration of actors is faced with several challenges, e.g. lack of consensus, a poor grasp of the problem, ignored alternatives, groupthink, conflicts, digressions, distractions, hidden agendas, poor planning, wrong people, poorly defined goals, premature decisions, lack of focus, misunderstandings, fear of speaking, and waiting to speak while others are dominating [23]. These are the challenges one would certainly expect when executing enterprise architecture creation as a collaborative problem solving task.

However, despite the above difficulties, collaboration is still essential for solving complex problems since no single individual possesses all the prerequisites (i.e. experience, resources, information) for problem solving [3,23]. Several technologies are in place to support collaborative problem solving or collaborative work in general, e.g. Group Support Systems (GSSs), web conferencing, virtual work spaces, teleconferencing, videoconferencing, dataconferencing, web-based collaboration tools, e-mail, and proprietary groupware tools [3,25]. This paper aims to demonstrate how GSSs can be used to support collaborative problem solving in enterprise architecture creation.

Moreover, the paper also discusses the design and evaluation of a method that is being developed using the Design Science research methodology, to complement enterprise architecture approaches with GSS functionalities (and support for collaborative problem solving). Design Science is a research paradigm that is used to develop innovative artifacts (i.e. processes, methods, models, frameworks etc) that offer solutions to significant problems in industry [10]. This implies that Design Science encourages practice-driven research since according to Hevner et al. [9,10], problems encountered in the business environment (or in practice) are treated as the requirements of any Information Systems research (in academia) that is conducted using Design Science. This is why this methodology is suitable for this research. The evolving method focuses on supporting Collaborative Evaluation of (Enterprise) Architecture Design Alternatives (CEADA). The method is therefore referred to as CEADA, pronounced as 'Keda'. This artifact draws from enterprise architecture approaches used in industry and collaborative problem solving theories developed in academia.

The CEADA method was initially evaluated using an analytical approach (see [20,21]). It has been further evaluated using an experiment and two real cases. Findings from this evaluation have been used to refine the method. This paper reports these findings and the refined models that describe CEADA. The findings indicate that the relevance and effectiveness of academia-based artifacts in addressing problems encountered in practice, can only be achieved through continuous and diverse evaluation of these artifacts in practice.

The remainder of this paper is structured as follows. Section 2 discusses the need for collaborative problem solving in architecture creation, while section 3 discusses the extent to which GSSs can be used to address this need. Section 4 explains how Design Science is used in this research, while section 5 presents the design of CEADA before it was evaluated. Section 6 discusses the evaluation of CEADA using an experiment and 2 real cases, and presents the refined CEADA models, while section 7 concludes the paper.

2   Collaborative Problem Solving in Architecture Creation

Despite the numerous benefits of enterprise architecture, its value proposition and the role of an enterprise architect are not understood in organizations accustomed to reactive decision making [13]. Program managers of such organizations (or who are used to independently devising mission-specific solutions) perceive enterprise architecture as a *"hostile takeover"* and may resist its creation, for fear of the new language and planning processes associated with it [2]. However, it has been reported that involvement of organizational stakeholders during architecture creation, to ensure that their concerns are considered, helps to create stakeholders' commitment [12]. It has also been reported that increasing stakeholders' involvement in the architecture creation process implies increasing their control in the process, which along with strong executive sponsorship can overcome resistances of architecture creation [2].

Therefore, it is likely that co-creation of enterprise architecture (i.e. having architects and organizational stakeholders collaboratively define and specify the enterprise architecture without implementing it) is likely to positively influence the success rate of implementing the specified architecture. Although we take this assumption to be true, this research does not involve studying the longer term impact of co-creation on the success rate of the implementation of the architecture. Rather, it involves studying effective ways of achieving architecture co-creation, where we suppose that proper stakeholders' involvement in architecture creation (i.e. co-creation or creative/collaborative problem solving) can be achieved through effective and efficient collaboration between stakeholders and architects. This implies that collaboration is a core thread in enterprise architecture development. Therefore, enterprise architects need to have a standard way of successfully managing their collaboration with stakeholders, even in the absence of a professional facilitator. Thus, increasing stakeholders' involvement in architecture creation certainly demands for amendments in enterprise architecture approaches. It has even been advised that in architecture creation, in addition to choosing a suitable enterprise architecture framework/approach, there is need to choose supporting methods and techniques [31,35] for e.g. enabling collaborative problem solving involved in architecture creation.

This research is therefore motivated to offer enterprise architects with an approach that can be used to increase stakeholders' involvement (and control) during architecture creation, by enabling effective and efficient collaborative problem solving. Although this vision is yet to be fully achieved through continuous validation of our method, it indicates the relevance of this research in practice. Moreover, proper stakeholder involvement and getting more acceptance of architecture results, are the key drivers of this research.

3   Group Support Systems in Architecture Creation

A GSS is an "interactive computer-based environment which supports concerted and coordinated team effort towards completion of joint tasks" [23]. GSSs include: Problem Structured Methods (PSMs, also known as model-based traditions or model-driven approaches); and Electronic Meeting Systems (EMSs, also known as workstation approaches or technology based or technology-driven approaches) [30]. A PSM enables one to represent a given situation using a model(s) so that participants can be able "to clarify their predicament, converge on a potentially actionable mutual problem or issue within it, and agree commitments that will at least partially resolve it" [17]. On the other hand, an EMS supports task-oriented collaborative work in (face-to-face) meeting processes that involve problem solving, decision making, deliberation, generating alternatives, negotiation, consensus building, and planning [25].

Although PSMs focus on understanding a given problem context from the perspective of participants, a skilled facilitator has a mandatory role and evaluating the performance of PSMs is difficult because their support varies depending on the uniqueness of the situation at hand [30]. Yet collaborative problem solving (or collaborative work in general) may consist of a combination of several (unique but interrelated) meeting processes, and therefore requires support that is flexible enough to quickly and efficiently facilitate any process [25]. For example, the nature of collaborative problem solving involved in enterprise architecture creation varies across organizations, but involves all the types of meeting processes listed above. This calls for flexible facilitation support which can be offered by EMS technologies (e.g. GroupSystems, MeetingWorks, TeamFocus, VisionQuest, and Facilitate.com), since they are equipped with capabilities for increasing effectiveness and efficiency of, and user satisfaction with, group meetings [25]. This implies that the nature of collaborative problem solving in enterprise architecture creation can be best supported by EMSs.

However, EMSs or GSSs in general have not been widely adopted by organizations, despite their numerous benefits [4,25]. This is mainly because GSSs have a high conceptual load (i.e. one has to put in a lot of effort to understand the intended effect of GSS functionalities for the user), and so organizations resort to hiring or training professional facilitators in order to be able to successfully use the technology [4]. A sustainable way that enables organizations to benefit from GSSs functionalities is collaboration engineering, which involves developing collaborative processes that can be used to support recurring mission-critical tasks and can be executed by practitioners themselves [5,4]. Therefore, since this research focuses on achieving successful collaborative problem solving in architecture creation without overdependence on professional facilitators, collaboration engineering is the suitable approach to benefiting from GSS functionalities during architecture creation. Sections 4 and 5 explain the approach that is being used to achieve this.

4   Design Science Research Methodology

Design Science guides the creation of innovative artifacts (e.g. methods, processes, or models that are relevant to a given application domain) using existing scientific knowledge (i.e. frameworks, theories, methods etc), and the evaluation of those artifacts using observational, analytical, experimental, descriptive, and testing methods [10]. According to Hevner [9], Design Science begins with identifying problems in, or opportunities for improving, the application domain. The application domain therefore initiates research by providing business needs (or research requirements) and acceptance criteria that are used to evaluate the resultant artifact [10].

In this research, the identified problem in the application domain (as earlier reported in [19,20]) was the challenge of effectively supporting collaborative problem solving or decision making during enterprise architecture creation (see top left part of Fig. 1). Moreover, as Fig. 1 shows, the evolving artifact to address this problem (or business need) is the CEADA method. The contents of the theoretical knowledge base (i.e. scientific theories, frameworks, models, and

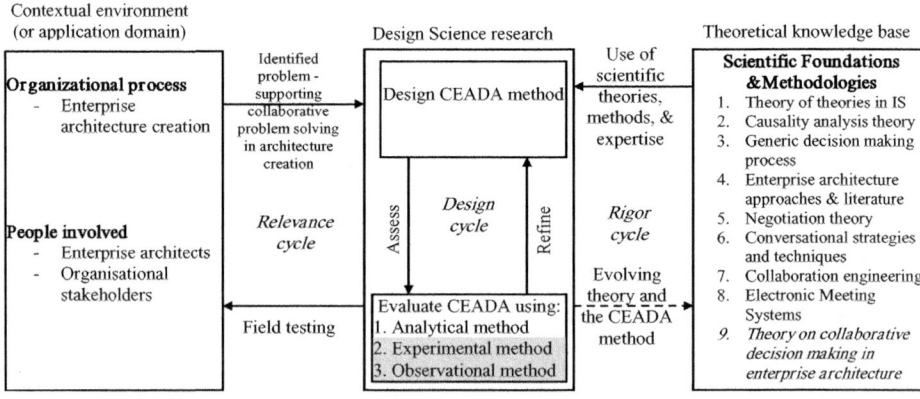

Fig. 1. Adapted Methodology - Design Science (based on [10] and [9])

methods) that were used to design CEADA are shown in the right part of Fig. 1. A discussion of how the contents of the theoretical knowledge base were adapted to design the CEADA method was earlier presented in [19,20,21,22], however section 5 gives a brief explanation of how the adaptation was done.

As shown in the middle part of Fig. 1, CEADA was designed and was first evaluated using an analytical method (i.e. structured walkthroughs) and findings were used to refine its design (see [20,21]). The design of CEADA was further evaluated using the experimental method (where a controlled experiment was used) and the observational method (where two real cases were used). This paper particularly reports the experimental and observational evaluation of CEADA.

According to Hevner et al. [10], observational design evaluation methods are: Case Study (i.e. conducting an in depth study of the designed artifact in a real business context); and Field Study (i.e. monitoring the use of the designed artifact in several projects). In this research, the Field Study method was used. Field Study evaluation of an artifact can be done using the action research method [9]. Action research, according to Susman and Evered [34], involves the following steps:

1. Diagnosing, i.e. identifying the main problem that is the root cause of the desire for change in an organization;
2. Action planning, i.e. specifying organizational actions that will address the main problem;
3. Action taking, which involves researchers collaborating with practitioners to implement the planned action so as to realize the desired changes in the organization;
4. Evaluating, which involves researchers and practitioners determining whether the theoretical effects of the action taken were realized;
5. Specifying learning, i.e. directing knowledge gained from the research (whether it was successful or not) to improve a theoretical framework or the organization's situation.

In action research, researchers actively participate with practitioners in the enquiry and change experiences involved in the research [1]. Since this was the first observational evaluation of CEADA, it was vital for the researchers to be actively involved in executing the method, before it could be evaluated in a setting where only practitioners are in charge of executing it. Action research was therefore the appropriate approach for undertaking the Field Study evaluation of CEADA. Details of how the above steps of action research were performed in this research are presented in section 6.3.

5   The CEADA Method

This section presents CEADA, its objectives, its design (i.e. components that address each objective), and its added value to the architecture approach. CEADA aims to enable collaborative problem solving to be successfully realized during enterprise architecture creation, even in the absence of a professional facilitator. It is designed using scientific knowledge i.e., the generic decision making theory [32], collaborative decision making (or negotiation) theory [28], the theory of theories in IS, casuality analysis theory [7], collaboration engineering [4,14], conversation strategies and techniques [26], enterprise architecture frameworks (particularly TOGAF [35]), literature on enterprise architecture creation, and the evolving theory on collaborative decision making in enterprise architecture creation [22]. An overview of how these theories apply to this research is given below.

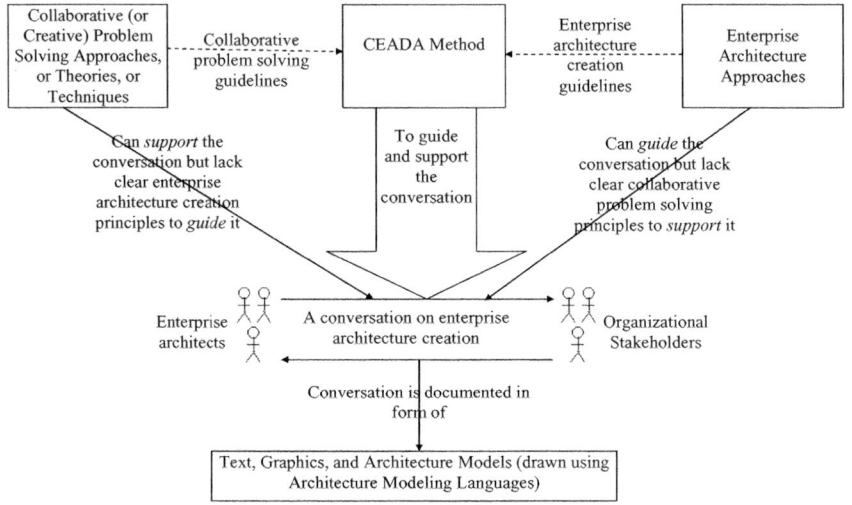

Fig. 2. The Added Value of CEADA to the Architecture Creation Methodology

Effective communication is essential for successful architecture creation among actors (i.e. stakeholders and enterprise architects) [22]. This communication can be perceived as a conversation [26], which in Fig. 2 we refer to as the architecture creation conversation. This conversation revolves about problem solving or decision making (as explained in section 2). Therefore, the conversation needs to be supported by the generic decision making process in [32]. Decision making in this conversation is collaborative in nature, since it includes stakeholders and architects. Therefore, the conversation needs to be supported by the collaborative decision making (or negotiation) theory in [28]. According to Simon [32] decision making involves studying the environment to identify the need for improvement/intervention (i.e. intelligence phase), devising possible decision alternatives (i.e. design phase), and choosing the most appropriate decision alternative (i.e. choice phase). Choosing a decision alternative involves assessing the possible decision alternatives and negotiating to agree on the most appropriate one. However, for negotiations to be successful there is need for effective collaboration among actors, which in turn creates a shared understanding among actors on the key issues in the conversation [22].

Fig. 2 justifies the need for a method that can: (1) support the conversation on enterprise architecture creation using collaborative decision making guidelines,

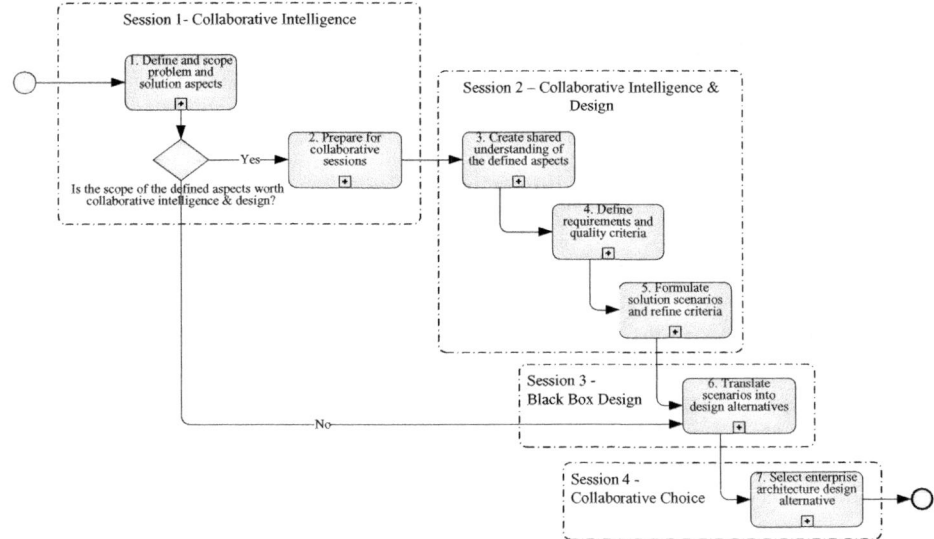

Fig. 3. Structure of the Architecture Creation Conversation

and (2) structure or guide the conversation using enterprise architecture creation guidelines. The latter are defined by The Open Group Architecture Framework (TOGAF) in [35] and the former can be realized by adapting (collaborative) decision making and other theories (see right part of figure 1) to suit architecture creation. Details of how this adaptation was done, to yield Fig. 3, were reported in [20,21,22], but in this section only a summary is given.

Fig. 3 shows the structural flow of the architecture creation conversation (which can also be perceived as the steps in, and requirements for, CEADA). It shows that the architecture creation conversation is divided into the following sessions.

1. *Collaborative intelligence session*, an adaptation of the intelligence phase (defined above) of the generic decision making process defined by Simon in [32]. It involves steps 1 and 2, i.e. define and scope problem and solution aspects and prepare for collaborative sessions with other stakeholders.
2. *Collaborative intelligence and design session*, an adaptation of Simon's intelligence and design phases. It involves steps 3, 4, and 5, i.e. create a shared understanding of the problem and solution aspects, define requirements and quality criteria, and formulate solution scenarios for the architecture.
3. *Black box design session*, an adaptation of Simon's design phase. It is essentially expert driven (involving enterprise architects only) and involves step 6, i.e. translate scenarios into enterprise architecture design alternatives.
4. *Collaborative choice session*, an adaptation of Simon's choice phase. It involves step 7, i.e. select a suitable (i.e. feasible, appropriate, and efficient) enterprise architecture design alternative. We consider an enterprise

architecture (or its design alternative) to be: appropriate if it is capable of addressing its planned purpose and realizing organization objectives; efficient if it addresses all stakeholders' concerns [24]; and feasible if it is achievable given the organization's resources.

The steps in Fig. 3 were decomposed to obtain the column labeled "activity description" in the design of CEADA that is shown in tables 1, 2, and 3. Collaboration engineering was then used to support the execution of the decomposed activities in the conversation.

Collaboration engineering approach (which was defined in section 3) introduces thinkLets, as building blocks for processes that can be executed by practitioners (in this case enterprise architects) to effectively manage collaborative recurring tasks, even in the absence of professional facilitators [14]. A thinkLet creates a pattern of thinking among people working toward a goal [5]. These patterns of thinking (or collaboration), according to Briggs et al. [5], include the following. (1) Generate (enables participants to move from having fewer concepts to more concepts that are shared by the group); (2) Reduce (enables participants to move from having many concepts to focus on fewer concepts that the group considers worthy of further attention); (3) Clarify (enables participants to move from having less to more shared understanding of concepts and phrases used to express them); (4) Organize (enables participants to move from less to more understanding of the relationships among concepts the group is considering); (5) Evaluate (enables participants to move from less to more understanding of the relative value of the concepts under consideration); and (6) Build consensus

Table 1. Design of Session 1 of CEADA

#	Activity Description	Deliverable(s)	Pattern of collaboration	ThinkLet(s)
\multicolumn{5}{l}{Session 1: Define and scope organization's problem and solution aspects (collaborative intelligence session)}				
1.1	Communicate purpose of the session and kind of information required from session	Guiding information	-	No ThinkLet required
1.2	Define basic information on business strategy, business objectives, and business requirements	Awareness of business strategy, objectives, and requirements	Generate, Reduce, Clarify	DealersChoice, FastFocus
1.3	Define organization's problem scope	Organization's problem scope		
	1.3.1 Identify aspects on the problem & its scope		Generate, Organize	OnePage, Concentration
	1.3.2 Agree on aspects of the problem & its scope		Build consensus	MoodRing
1.4	Identify external solution constraints (from e.g. regulatory authorities)	External constraints	Generate, Clarify	OnePage, FastFocus
1.5	Define purpose of the architecture effort	Purpose of the architecture effort		
	1.5.1 Generate ideas on purpose of architecture effort		Generate, Organize	OnePage, Concentration
	1.5.2 Agree on purpose of architecture effort		Build consensus	MoodRing
1.6	Define high level solution specifications	General solution specifications		
	1.6.1 Generate ideas on solution specifications		Generate	FreeBrainstorm
	1.6.2 Filer generated solution specifications		Reduce, Clarify	FastFocus
	1.6.3 Agree on solution specifications		Evaluate	StrawPoll, CrowBar
1.7	Seek shared understanding on the scope of the problem and its solution, and seek consensus on whether the scope of these aspects is worth a collaboration effort of organization key stakeholders	Understanding scope of problem and its solution, and appreciation of need for collaborative effort	Build Consensus	MoodRing
1.8	Select key stakeholders to participate in subsequent collaboration efforts with enterprise architects (and define their roles)	Other key stakeholders to collaborate with enterprise architects	Generate	No ThinkLet required
1.9	Reveal calendar of events, communicate the expectations of architect team, and find out stakeholders' expectations in the subsequent collaboration efforts during the architecture effort	Calendar of events and expectations of architects and stakeholders	-	

Table 2. Design of Session 2 of CEADA

#	Activity Description	Deliverable(s)	Pattern of collaboration	ThinkLet(s)
Session 2: Seek shared understanding of problem and solution aspects, and define requirements & quality criteria (collaborative intelligence and design session)				
2.1	Communicate the purpose of the session and kind of information required	Guiding information	-	No ThinkLet required
2.2	Stakeholders share their concerns about the problem and solution aspects	Stakeholders' concerns	Generate	LeafHopper
2.3	Categorize concerns by type and organization domains	Categories of stakeholders' concerns	Reduce, Clarify	Popcorn sort
2.4	Analyze and discuss concerns while seeking a shared understanding of the problem and solution aspects	Shared understanding of problem & solution aspects, refined concerns	Organize	BucketWalk, BucketBriefing
2.5	Validate stakeholders' concerns	Valid concerns	Evaluate	StrawPoll
2.6	Agree on amendments to problem and solution aspects (i.e. the as-is and to-be situation)	Refined problem and solution aspects	Build Consensus	StrawPoll, Red-Light-Green-Light
2.7	Brainstorm on requirements, based on valid stakeholder's concerns, that the architecture must address	Requirements for the architecture	Generate	Free-Brainstorm
2.8	Validate requirements for the architecture	Valid requirements	Reduce, Clarify, Organize	Popcorn sort
2.9	Agree on requirements for the architecture	Consensus on architecture requirements	Evaluate, Build Consensus	StrawPoll, BucketWalk
2.10	Brainstorm on business, governance, & operational quality criteria for evaluating design alternatives	Business, governance, & operational quality criteria	Generate	Free-Brainstorm
2.11	Validate quality criteria	Valid quality criteria	Reduce, Clarify, Organize	Popcorn sort
2.12	Agree on (business, governance, & operational) quality criteria	Consensus on quality criteria	Evaluate, Build Consensus	StrawPoll, BucketWalk
Session Break				
2.13	Communicate purpose of session and kind of information required	Guiding information	-	No ThinkLet required
2.14	Brainstorm on types of solution scenarios to be formulated	Required types of solution scenarios	Generate	Free-Brainstorm
2.15	Identify components of a solution scenario	Components of solution scenarios	Generate	Comparative Brainstorm
2.16	Assemble components of solution scenarios	Solution scenarios	Generate, Organise	Could-Be-Should-Be, BranchBuilder
2.17	Refine (business, governance, & operational) quality criteria	Detailed quality criteria	Clarify, Build Consensus	BucketWalk, Red-Light-Green-Light

Table 3. Design of Sessions 3 and 4 of CEADA

#	Activity Description	Deliverable(s)	Pattern of collaboration	ThinkLet(s)
Session 3: Translate solution scenarios into architecture design alternatives (black box design session)				
Session 4: Select suitable enterprise architecture design alternative (collaborative choice session)				
4.1	Communicate purpose of session and kind of information required	Guiding information	-	No ThinkLet required
4.2	Explain positive and negative implications of analyzed design alternatives to stakeholders	Positive and negative implications of the enterprise architecture design alternatives	-	
4.3	Seek shared understanding (among stakeholders) on the implications of the analyzed design alternatives	Shared understanding on relevant information for making the final decision	Evaluate	StrawPoll, CrowBar
4.4	Select feasible, appropriate, & efficient design alternative (using the quality criteria from sessions 1 and 2)	Consensus on feasible, appropriate, & efficient design alternative	Evaluate, Build Consensus	MultiCriteria, Red-Light-Green-Light

(enables participants to move from having fewer to more group members willing to commit to a proposal).

According to Kolfschoten and Vreede [14], a collaboration processes (that enables participants to undergo the above patterns of thinking) is designed using the following procedure. (1) Task diagnosis (which involves defining the goal and deliverables of the collaboration process); (2) task decomposition (which involves defining the basic activities for achieving the defined goal and deliverables); (3) ThinkLet choice (which involves using some criteria to assign each basic activity a suitable thinkLet that will guide its completion); (4) agenda building (which involves assembling the activities and their corresponding patterns of thinking as well as thinkLets so as to validate the process); (5) design validation and evaluation; and (6) documentation. In [20], it is reported how this procedure was applied in this research to obtain the design of CEADA presented in tables 1, 2, and 3.

6   Performance Evaluation of CEADA

This section reports the evaluation of the design and performance of CEADA in an experimental setting and in a real business setting. In this evaluation, CEADA was used along with TOGAF, Business Process Modeling Notation (BPMN), and ArchiMate modeling concepts to create only the architecture vision for each case.

6.1   Criteria for Evaluating the Performance of CEADA

The criteria for evaluating the performance of CEADA were derived from: the theory of collaborative decision making into architecture creation (see [22]); the requirements for deploying collaborative decision making into architecture creation (see [21]); and from the issues discussed in sections 2 and 3. The criteria are classified into effectiveness and efficiency. Effectiveness in this case refers to the ability of CEADA to support the following.

1. Creation of a shared understanding of the organization's problem and solution aspects among stakeholders and architects;
2. Creation of stakeholders' commitment towards the success of architecture creation;
3. Explicit description and agreement on the requirements, quality criteria, and solution scenarios that the architecture must address; and
4. Selection and agreement on a suitable enterprise architecture design alternative.

Efficiency criterion in this case refers to the ability of CEADA to satisfy criteria 1 - 4 above in the shortest possible time. The performance of CEADA under criteria 1 – 4 above was measured by the following indicators.

1. Shared understanding among stakeholders was measured by the level of consensus among stakeholders on concerns and requirements that the architecture must address;
2. Stakeholders' commitment was measured by stakeholders' dedication to accomplishing the activities in the CEADA method;
3. Agreement on requirements, quality criteria, and solution scenarios was measured by the level of consensus among stakeholders on these;
4. Selection of (and agreement on) the suitable design alternative was measured by the level of consensus on a chosen design alternative.

The level of consensus (in indicators 1, 3, and 4 above) was measured by the standard deviation of the priorities or weights that stakeholders assign to the items of interest in a given session. Data on the evaluation of CEADA was gathered using questionnaires, observation, and GSS data logs. Questionnaires were filled by all participants and observation of the execution environment of CEADA was done by the researchers. Meetingworks^{TM} was the GSS technology that was used to support the execution of CEADA. Stakeholders' dedication in indicator 2 above was measured by their attendance, participation, and enthusiasm in the collaborative sessions.

6.2   Experimental Evaluation of CEADA

In Design Science, prior to evaluating an artifact using real case studies, an experimental evaluation of the artifact is vital [11]. CEADA therefore was experimentally evaluated before it was evaluated using real cases. Experimental evaluation involves studying the usability qualities of a designed artifact in a controlled environment, and executing it with artificial data [10]. As discussed in section 4, the experimental evaluation of CEADA was implemented using action research.

In the experiment, the following steps of action research defined by Susman and Evered [34] (see section 4) were undertaken. At diagnosing step, a fictitious organization was chosen, whose main challenge was implementing its strategy of expanding from a national University to a networked European University. At action planning step, it was determined that the national university had to develop an enterprise architecture for the networked European university. The enterprise architecture would then guide and inform the transformation from a national university to a networked European university. Thus, the purpose of CEADA in experiment was to support collaborative creation of the enterprise architecture vision of the networked European university. At Action taking step, CEADA was used to support the architecture creation conversation in experiment. At evaluating step, the design and performance of CEADA were evaluated by the participants (who played the role of stakeholders in the national university) and the researchers. At the step of specifying learning, lessons learned from this evaluation have been used to improve CEADA.

Experiment Setup and Execution. In the experiment an example case was used and participants were 26 students undertaking the course of Information Architecture at Radboud University Nijmegen (The Netherlands). The experiment theme was to create an (enterprise) architecture for the education and examination institute of a networked European university. The architecture of this institute was to include the required business/operational processes, data flows, application systems, and technology infrastructure. Participants were divided into enterprise architects and stakeholders. The stakeholders were further divided into 6 groups, where each group took up any of the following roles: director, educational coordinator, lecturer, administrative staff, IT technical staff, and the students' representative.

Three collaborative sessions, each with a duration of 2 hours, were conducted supported by the design of CEADA shown in tables 2 and 3. The first session aimed at enabling participants to acquire a shared understanding of the problem and solution aspects involved in creating the architecture of the institute; and define the concerns, requirements, and quality criteria that must be addressed by the architecture. The second session aimed at enabling participants to formulate solution scenarios that the architecture must address. These scenarios were then used by participants playing the architects' role to create three possible architecture design alternatives. The third session aimed at enabling participants to select and agree on the suitable architecture design alternative of the institute. In the three sessions researchers played the role of facilitator and observer.

Results From the Experiment. The results in table 4 were obtained from the questionnaires that were used to gather data on the evaluation of the performance of CEADA in the experiment. This questionnaire survey approach to measuring participants' satisfaction with a collaboration process and its outcome was introduced by Briggs et al. [3]. In these questionnaires, we used the 5 point Likert scale questions, with responses ranging from strongly disagree (point 1) to strongly agree (point 5).

Lessons Learned From the Experiment. During the collaborative sessions all participants playing the stakeholders' role worked in one group when executing activities shown in tables 2 and 3. This immensely affected the level of consensus among stakeholders on: the requirements and solution scenarios that the architecture had to address; and on the suitable architecture design alternative. Moreover, from the questionnaires filled by participants it was noted that some stakeholders did not understand how their concerns and requirements were catered for in the three architecture design alternatives that the architects had designed (see table 4). Other stakeholders felt that their concerns were not addressed at all.

On reflecting upon how these issues could have been avoided, it was noted that when executing some activities in the collaborative sessions, participants or stakeholders would have been divided to work in small groups formed based on their specialization area. The activities that required stakeholders to be divided in small groups are 2.2, 2.3, 2.7, 2.8, 2.10, 2.11, 2.14 – 2.16, and 4.2 – 4.4

Table 4. Performance Evaluation of CEADA in the Experiment

#	Evaluation Criteria for CEADA	Indicator	
		Mean score	Standard deviation of scores
1	Satisfaction with the activities done in the collaborative sessions	2.00	0.88
2	Satisfaction with the outcome(s) of the collaborative sessions	2.05	0.91
3	Collaborative sessions helped to increase understanding of the concerns and requirements of all units in the organisation	3.89	0.94
4	Collaborative sessions helped stakeholders to freely express their views about the current operations in the organisation	3.53	1.22
5	Collaborative sessions helped stakeholders to understand why some of their concerns/views were not chosen/voted by others during the sessions	3.11	1.05

(see tables 2 and 3). This is because it was noted that stakeholders from a given specialization/unit would assign high priorities to concerns and requirements that pertain to their unit and then assign low priorities to those from other units. This is why when evaluating architecture requirements and design alternatives, results indicated that there was a low level of consensus on the concerns, requirements, and the design alternative that was chosen. Thus, the division of stakeholders into small groups during the execution of these activities will enable them to explicitly define and quickly reach consensus on the requirements of a given unit. Moreover, during the selection of architecture design alternatives (i.e. activities 4.2 – 4.4 in table 3), there was a need for architects to first explain the architecture to the small groups of stakeholders such that each stakeholder sub group can gain a shared understanding of how their concerns are addressed in the architecture viewpoint that pertains to them.

Furthermore, it was observed that at the completion of an activity that requires division of stakeholders, they can all meet together (in a short plenary session) to identify any overlapping requirements or ambiguities; and to acquire an understanding of requirements from other units or stakeholder subgroups. Lastly, activity 2.6 (see table 2) was a repetition of activity 2.5, it was therefore deleted. These lessons learned from the experiment were used to refine the design of CEADA, which was further evaluated as discussed below.

6.3 Field Study Evaluation of CEADA

As discussed in section 4, Field Study as a design evaluation method was used, and was implemented using action research. Since Case Study can be used to describe a unit of analysis (such as an organization) or a qualitative research method [18], Case Study as used in this section refers to the organization in which CEADA was evaluated, but not a qualitative research method.

Case Study 1: Nsambya Home Care (NHC). This is a donor funded organization whose mission is to offer free services to HIV positive patients in Uganda. It has the following units.

1. Medical unit, which is divided into the HIV medical unit – that clinically monitors HIV positive patients; and the Tuberculosis (TB) unit – a referral TB unit in Uganda, that treats TB patients and finds out how many of them are actually HIV positive.
2. Pharmacy unit, which dispenses prescribed drugs to patients and manages stock and orders of drugs.
3. Laboratory unit, which monitors laboratory investigations for patients.
4. Psychosocial unit, which manages relations between NHC and its patients, listens to patients' social and psychological issues, counsels, and sensitizes patients on the do's and don'ts of HIV.
5. Finance and administration unit, which manages incomes and expenditures, and oversees pharmacy, laboratory, and psychosocial units.

6. Monitoring and evaluation (or data) unit, which assembles and tracks all activities in NHC, collects reports from all units, compiles them and sends them to the right destinations. This unit reports to the assistant coordinator of NHC, who oversees the implementation of planned activities. NHC currently has a LAN which has 3 data servers and a few computers that are used in the pharmacy, laboratory, finance, cash office, and data units. The computers are mainly networked for Internet usage only.

In NHC, the following steps of action research defined by Susman and Evered [34] (see section 4) were undertaken. At diagnosing step, it was discovered that the main challenge NHC was facing is the hectic and time consuming process of capturing and retrieving records or data when compiling reports for the donors. At action planning step, it was determined that NHC has to refine its operational flow in order to ensure effective and efficient data capturing, retrieval, sharing, storage, and reporting. The best way to achieve this was through developing an enterprise architecture (vision) that would guide and inform the desired transformation in NHC. Consequently, the purpose of CEADA in NHC was to support collaborative creation of the enterprise architecture vision of NHC. At Action taking step, CEADA was used to support the architecture creation conversation in NHC. At evaluating step, the design and performance of CEADA were evaluated by the stakeholders and the researchers. The effects of the architecture that was created will only be determined after the architecture is implemented. However, architecture implementation is beyond the scope of this research. At the step of specifying learning, lessons learned from this evaluation have been used to improve CEADA. Moreover, after the architecture that was created is implemented, NHC's problematic situation will be addressed.

Results From Case Study 1. CEADA was used to support the architecture creation conversation in NHC, which involved 13 stakeholders (where 5 were from the data unit, 5 were from the pharmacy unit, and 3 were from the psychosocial unit). Table 5 summarizes the performance of CEADA under criteria 1, 3, and 4 of effectiveness that were explained in section 6.1.

The results in table 5 were obtained from the questionnaires that were used to gather data on the evaluation of the performance of CEADA in NHC. Like in the experiment, these questionnaires had the 5 point Likert scale questions, with responses ranging from strongly disagree (point 1) to strongly agree (point 5).

Table 5. Performance Evaluation of CEADA in NHC

#	Evaluation Criteria for CEADA	Indicator	
		Mean score	Standard deviation of scores
1	Satisfaction with the activities done in the collaborative sessions	4.20	0.42
2	Satisfaction with the outcome(s) of the collaborative sessions	4.20	0.42
3	Collaborative sessions helped to increase understanding of the concerns and requirements of all units in the organisation	4.50	0.53
4	Collaborative sessions helped stakeholders to freely express their views about the current operations in the organisation	4.50	0.53
5	Collaborative sessions helped stakeholders to understand why some of their concerns/views were not chosen/voted by others during the sessions	3.30	1.25

The business, data, applications and technology architecture models that constitute the selected design alternative for the architecture vision of NHC, and some
photos of the group sessions are shown in Fig. 4, 5, and 6.

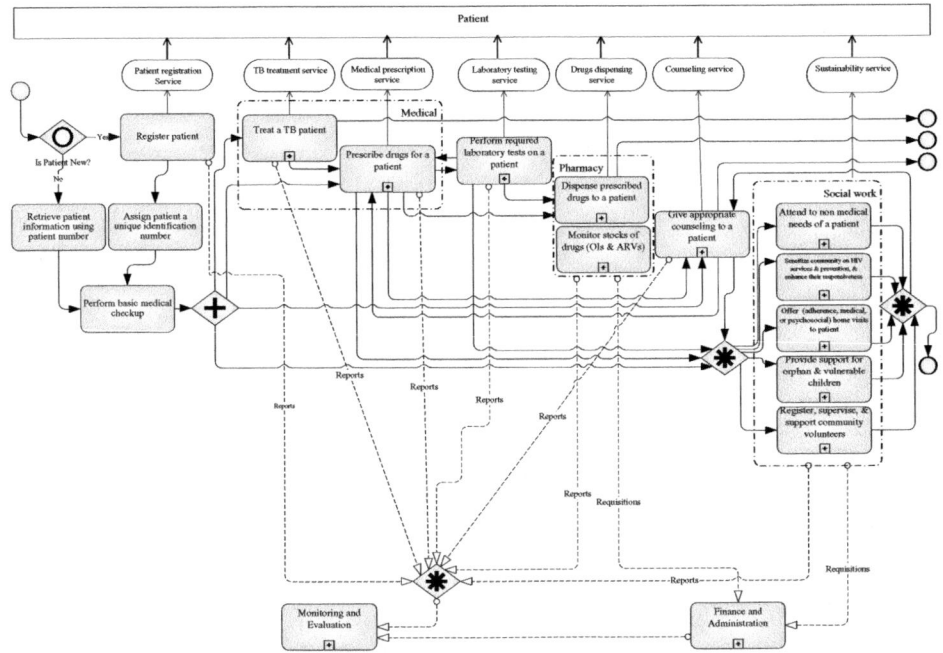

Fig. 4. Architecture Vision - Processes in NHC

Refinement of CEADA Using Lessons Learned From Case Study 1. In
NHC, activities 1.1 – 1.8 in session 1 of CEADA (see table 1) did not require support from a GSS, instead they were executed using interviews with an executive
member. This implies that, in practice, session 1 can be executed as a "collaborative intelligence session" or "intelligence interview session" (see figure 7).
The latter means that in the architecture creation conversation, one well informed stakeholder represents others (to define and scope the organization's
problem and desired solution), and the former means that several stakeholders
have to be involved in this session of the conversation. If session 1 has to be
executed as a collaborative intelligence session, then there is need for support
from a GSS. Note that if session 1 has to be executed as an intelligence interview
session, this does not affect the performance of CEADA, since the problem and
solution aspects defined in session 1 are refined and elaborated by all key stakeholders in session 2. Furthermore, the division of stakeholders into small groups
(as a lesson learned from the experiment) enabled session 2 to be successful,

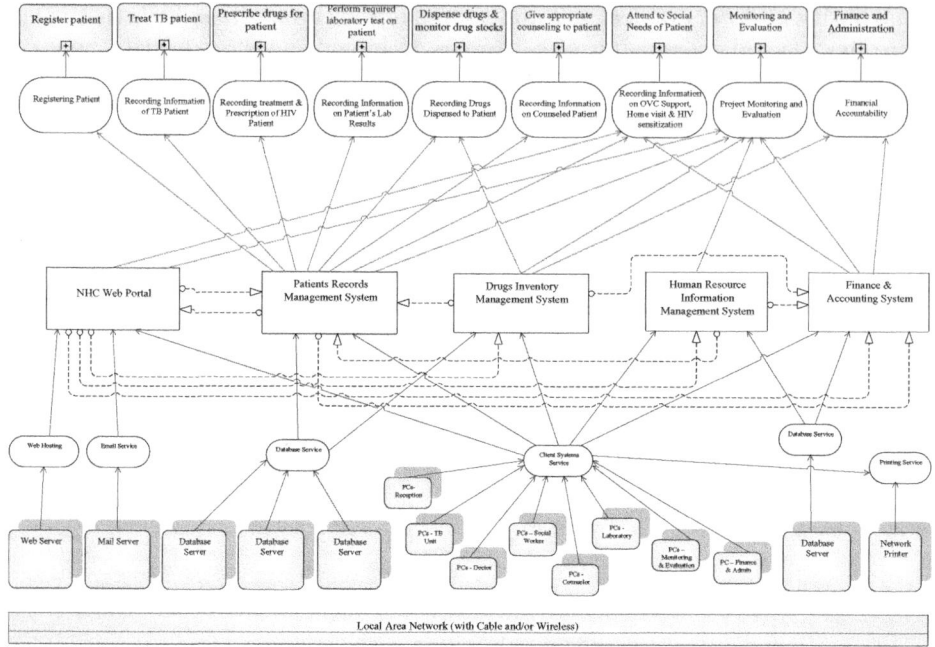

Fig. 5. Architecture Vision - Processes, Application Systems, and Technology in NHC

Fig. 6. Group Session Scenes

in the sense that stakeholders quickly reached a high level of consensus on the requirements that the architecture had to address.

In NHC, it was noted that enterprise architecture design alternatives can be divided into 2 levels, i.e. organization wide level and departmental/unit level architecture design alternatives. Organization wide architecture design alternatives involve considering, e.g., whether a given business process in a given department can be outsourced or not, or whether two/more departments or business processes can be merged into one. For NHC examples of organization wide architecture design alternatives include the following.

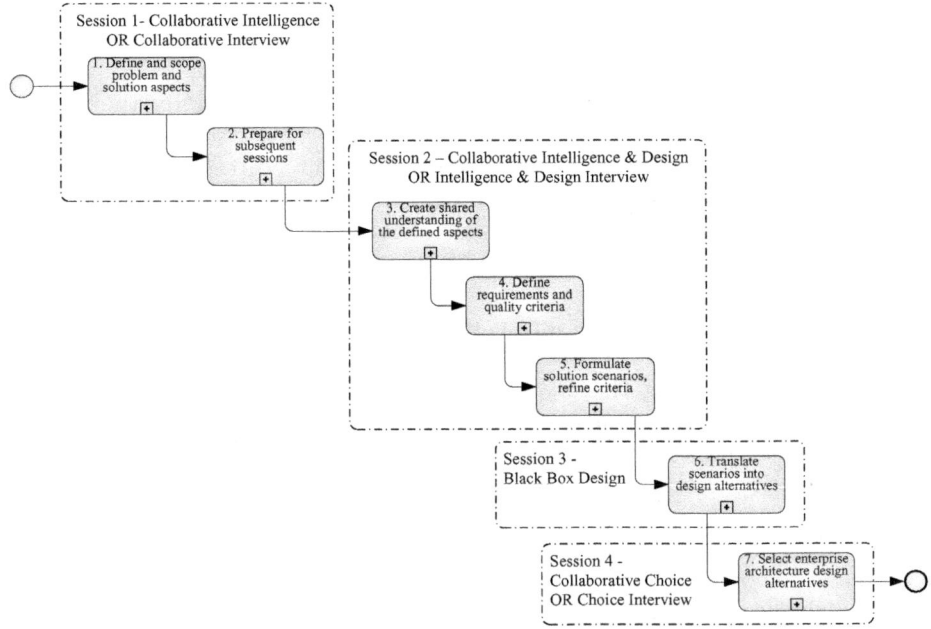

Fig. 7. Refined Structure of the Architecture Creation Conversation

1. Excluding the operational processes of the TB unit from the NHC architecture since the TB patients are treated separately from the HIV positive patients, in terms of medical prescriptions and geographical proximity of the TB unit to other NHC units.
2. Including all operational processes in NHC in the architecture (as shown in figures 4 and 5).
3. Avoiding the risk of unauthorized users hacking into the patient's records management system, by not connecting it to the NHC web portal. The limitation of this alternative was that the patient's records management system could then not be accessed by the staff who offer home visiting services or treatment to patients.
4. Connecting the patient's records management system to the web portal and ensuring high quality security and authentication measures (which definitely has financial implications).

Departmental/unit level architecture design alternatives represent different ways in which things can be done within a given unit to achieve effectiveness. The organization board may not be very relevant at this level, provided the chosen way of operation in a given unit is within the policies of the organization. For NHC, examples of unit level architecture design alternatives include the following. (1) Having a new or improved format of forms for capturing patient data into the patient records management system rather than the format of the existing paper based forms that are currently being frequently forwarded to the data unit for data entry. (2) Retaining the format of the existing paper based forms and simply using the same forms to capture patient data into the patient's records management system.

The two levels of design alternatives explained above, justify the need for stakeholders to be divided into small groups (based on their specialization) when formulating solution scenarios and evaluating architecture design alternatives (i.e. activities 2.14 – 2.16, and 4.2 – 4.4 in tables 2 and 3). Furthermore, the two levels of architecture design alternatives indicate that activities in session 4 (see table 3) have to be further decomposed so that evaluation of design alternatives is done at two levels.

It was also noted that when formulating solution scenarios (i.e. activities 2.14 – 2.16 in table 2), there is limited use of GSS in CEADA. What was required was more hands on and negotiation rather than GSS usage. This is because when executing these activities stakeholders don't see the need for punching their ideas into the GSS. They instead simply start sketching out what they mean rather than to describe it. The main use of GSS in activities 2.14 – 2.16 then remains to store comments or remarks made during these activities, so that they can be discussed by the group.

It was also noted that there is a need to identify a suitable negotiation model to improve the negotiation process required when executing activities 2.5, 2.8, 2.11, 2.14, 2.15, 4.4, and 4.5 (in tables 6 and 7) of the collaborative sessions. This is because thinkLets alone were not enough to fully support the negotiation required

Table 6. Refined Design of Session 2 of CEADA

#	Activity Description	Arrangement of stakeholders	Pattern of collaboration	ThinkLet(s)
	Session 2: Seek shared understanding of problem and solution aspects, and define requirements & quality criteria (collaborative intelligence and design session)			
2.1	Communicate the purpose of the session and kind of information required	All	-	-
2.2	Stakeholders share their concerns about the organizational problem and solution aspects	Divide based on specialization	Generate	LeafHopper
2.3	Categorize concerns by type and organization domains/units/departments	Divide based on specialization	Reduce, Clarify	Popcorn sort
2.4	Analyze and discuss concerns while seeking a shared understanding of all problem and solution aspects	All	Organize	BucketWalk, BucketBriefing
2.5	Validate stakeholders' concerns	All	Evaluate, Build Consensus. Need for in-depth negotiation	StrawPoll, Red-Light-Green-Light
2.6	Based on the valid stakeholders' concerns, brainstorm on (business) requirements that the architecture must address	Divide based on specialization	Generate	Free-Brainstorm
2.7	Categorize and discuss requirements for the architecture	Divide based on specialization	Reduce, Clarify, Organize	Popcorn sort
2.8	Validate and agree on requirements for the architecture	All	Evaluate, Build Consensus Need for in-depth negotiation	StrawPoll, BucketWalk
2.9	Brainstorm on business, governance, & operational quality criteria for evaluating design alternatives	Divide based on specialization	Generate	Free-Brainstorm
2.10	Categorize and discuss quality criteria	Divide based on specialization	Reduce, Clarify, Organize	Popcorn sort
2.11	Validate and agree on (business, governance, & operational) quality criteria	All	Evaluate, Build Consensus Need for in-depth negotiation	StrawPoll, BucketWalk
	Session Break			
2.12	Communicate purpose of session and kind of information required	All	-	-
2.13	Brainstorm, clarify, & agree on the possible organization wide solution scenarios that address the organization wide concerns and requirements	All	Generate, organize, Build Consensus Need for in-depth negotiation	Free-Brainstorm, Could-Be-Should-Be, BranchBuilder
2.14	Brainstorm, clarify, & agree on the possible departmental/unit level solution scenarios that address the unit specific concerns and requirements	Divide based on specialization	Generate, organize, Build Consensus Need for in-depth negotiation	Comparative Brainstorm, Could-Be-Should-Be, BranchBuilder
2.15	Evaluate all unit specific solution scenarios in context of the organization wide solution scenario that was chosen in 2.13	All	Organize, Build Consensus Need for in-depth negotiation	Could-Be-Should-Be, BranchBuilder
2.16	Refine (business, governance, & operational) quality criteria based on the formulated solution scenarios	All	Clarify, Build Consensus	BucketWalk, Red-Light-Green-Light

Table 7. Refined Design of Sessions 3 and 4 of CEADA

	Session 3: Translate solution scenarios into architecture design alternatives (black box design session)			
	Session 4: Select suitable enterprise architecture design alternative (collaborative choice session)			
#	Activity Description	Arrangement of stakeholders	Pattern of collaboration	ThinkLet(s)
4.1	Communicate purpose of session and kind of information required	All	-	-
4.2	Explain the architecture design alternatives for each unit specific solution scenario, and the positive and negative implications of each alternative, to only the group of stakeholders who are affected by a given solution scenario	Divide based on specialization	-	-
4.3	Seek shared understanding (among stakeholders in each small group/unit) of the implications of each architecture design alternative of each unit specific solution scenario	Divide based on specialization	Evaluate	StrawPoll, CrowBar
4.4	Choose a suitable architecture design alternative for each unit in the organization	Divide based on specialization	Evaluate, Build Consensus / Need for in-depth negotiation	MultiCriteria, Red-Light-Green-Light
4.5	Assess and discuss the compatibility of all the chosen architecture design alternatives for all unit specific solution scenarios	All	Need for in-depth negotiation	-
4.6	Choose the suitable (i.e. feasible, appropriate, and efficient) enterprise architecture design alternative	All	Evaluate, Build Consensus	MultiCriteria, Red-Light-Green-Light

when executing those mentioned activities. This explains why results in table 5 indicate that the majority of the stakeholders were not sure whether they understood why some of their concerns were not chosen by others during the collaborative sessions (see the last 3 rows in table 5). All findings from Case Study 1 were used to refine the design of CEADA as shown in Fig. 7 and tables 6 and 7.

Case Study 2: Makerere University Guest House (MUKGH). This offers hotel services to the Makerere University community, to the guests visiting the University, and to the general public. The mission of MUKGH is to offer the most distinguished and customer responsive services that ensure repeat customers and loyalty. The vision of MUKGH is to become the most preferred referral guest house in Kampala city (Uganda). MUKGH has defined strategic business objectives that it will strive to achieve in its efforts to serve its clientele. Its financial business objectives include: increasing gross revenue by 50% by 2011; increasing the profit margin to 25% by 2011; increasing revenues in the medium term plan by four fold by 2016; and increasing the room occupancy to 95%. Its service delivery business objectives include: improving the quality of its products and services to competitive standards; improving the efficiency and effectiveness in the operations of the business; upgrading to a 100 room 3 star hotel by 2016, automating its booking system; and implementing an internship program that will enable the university to tap into the large student pool.

Like in Case Study 1, the following steps of action research defined by Susman and Evered [34] were undertaken in MUKGH. At diagnosing step, it was discovered that the main challenge MUKGH was facing is the lack of the basic infrastructure and management to deliver the quality service desired by her clients. At action planning step, it was determined that MUKGH has to address this issue by tapping into its unexploited potential through pursuing the following strategic goals. (1) Upgrading to a 3-star hotel in order to provide quality services and increase its customer base. This will entail restructuring of the current facilities and the way of working, so as to meet the minimum industry requirements and market demands (2) Improving business efficiency and effectiveness through adoption of modern hospitality business management practices. (3) Product and service diversification. The best way for MUKGH to achieve its strategic goals was through developing an enterprise architecture (vision) that would guide and inform its desired transformations. Therefore, the purpose of CEADA in MUKGH was to support collaborative creation of the enterprise architecture vision of MUKGH. At Action taking step, CEADA was used to support the architecture creation conversation in MUKGH. At evaluating step, the design and performance of CEADA were evaluated by the stakeholder who participated in the conversation and the researchers. At the step of specifying learning, the lessons learned from the evaluating step have been used to refine CEADA. The problem in MUKGH will be addressed after the architecture that was created is implemented.

Results From Case Study 2. In MUKGH, session 1 of CEADA was also supported by interviews (where the manager represented all other key stakeholders) to define the problem and solution aspects. In other words, it was executed as an intelligence interview session (which is explained under lessons learned from Case Study 1). On completing session 1, a situational application of CEADA was encountered. This situational application is indicated by a gateway after step 1 of CEADA (see Fig. 3 in section 5). The "no" arrow in this gateway means that in a situation where the scope of the problem and solution aspects does not require a collaborative intelligence and design session, architects are to design the architecture in a black box session. However, for MUKGH, a collaborative intelligence and design session was not necessary not because of the scope of the problem and solution aspects, but due to political and operational issues that cannot be discussed here due to confidentiality reasons. Therefore, although collaborative intelligence and design may not be necessary, there is still need for architects to collaborate with at least one of the well informed stakeholders. Thus, there was need to refine the structure of the conversation to indicate that the "no" arrow in the gateway means that in some situations, there might not be need to execute session 2 as "collaborative intelligence and design", but rather as "intelligence and design interview session". Similarly, in some situations (like in MUKGH) session 4 can also be executed as a "choice interview session" rather than a "collaborative choice session". Fig. 7 shows these refinements in the structure of the collaborative architecture creation conversation. The business, data, applications and technology architecture models that constitute the selected design alternative for the enterprise architecture vision of MUKGH are shown in Fig. 8, 9, and 10.

Refinement of CEADA Using Lessons Learned From Case Study 2. The two levels of architecture design alternatives that were identified in Case Study 1 (i.e. organization wide architecture design alternatives and departmental level architecture design alternatives) were also identified in Case Study 2. This confirmed that there was need to cater for the two types of architecture design alternatives in the design of CEADA. This refinement was made in activities 2.13 – 2.15 in table 6. All findings from Case Study 2 (as discussed above) were used to refine the design of CEADA as shown in Fig. 7 and tables 6 and 7.

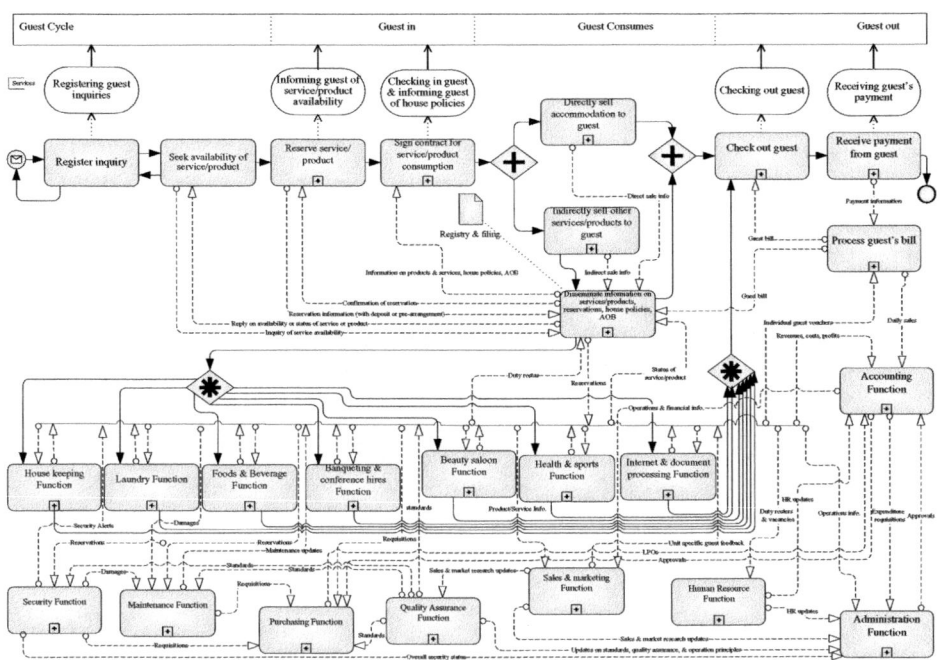

Fig. 8. Architecture Vision - Processes in MUKGH

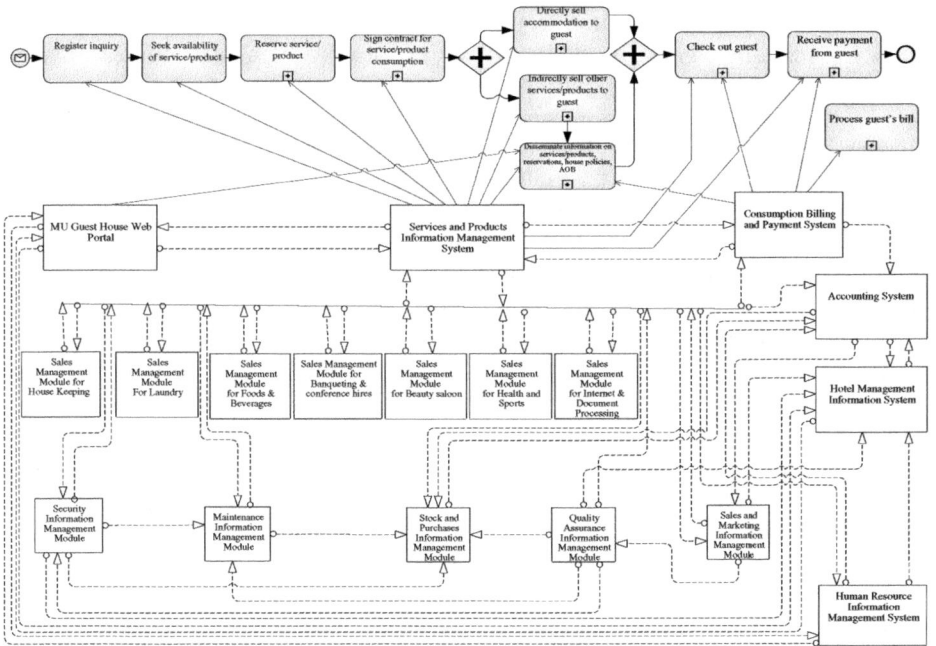

Fig. 9. Architecture Vision - Processes and Application Systems in MUKGH

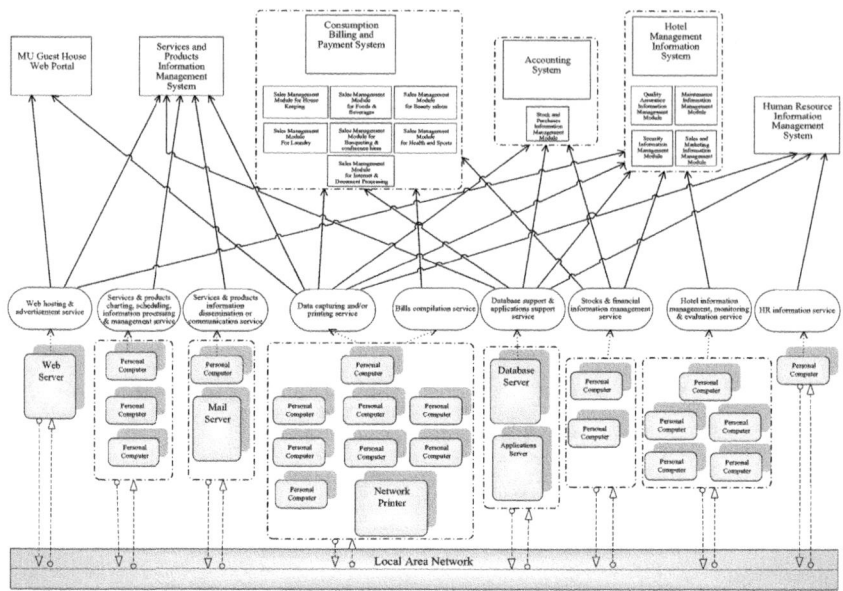

Fig. 10. Architecture Vision - Application Systems and Technology in MUKGH

7   Conclusions and Future Work

The evaluation of the design and performance of CEADA, using the experiment and Case Studies, generally helped to improve its design. Although the performance of CEADA in the experimental evaluation was not good, the experiment revealed a major weakness of the method. From the experimental evaluation, it was observed that it is vital to divide stakeholders into small groups (based on their area of specialization) when executing some activities. This observation was used to refine CEADA, and was tested in Case Study 1. The results from Case Study 1 indicated an improvement in the performance of CEADA. The results indicate that CEADA successfully supported the architecture creation conversation in NHC (i.e. Case Study 1).

Moreover in Case Study 1, two levels of enterprise architecture design alternatives were encountered i.e., organization wide architecture design alternatives and departmental level architecture design alternatives. The two levels of design alternatives were also encountered in Case Study 2. Consequently, the design of CEADA for session 2 of the architecture creation conversation was modified to cater for evaluation of the two levels of design alternatives. Case Study 2 was a situational application of CEADA, in which the architecture creation conversation was done using interviews rather than support from GSS. This was mainly due to the political and operational issues that were encountered in MUKGH. Therefore, the evaluation of CEADA specifically helped to: (1) identify the situational applications of CEADA; and (2) identify the weaknesses (and strengths) of CEADA, which were worked on to improve its design.

The future refinement of CEADA involves identifying and adapting a suitable negotiation model that will support negotiations required when executing some activities in the collaborative sessions of CEADA. Moreover, CEADA will be evaluated in parallel with other methods so as to compare its advantages and disadvantages in relation to other methods.

Acknowledgements. We are very grateful to Daniel Kalibbala (the data manager of NHC), Joseph Kayizzi (the manager of MUKGH), and the administration and staff members of NHC, for their participation in the collaborative sessions and their valuable contributions into this research.

References

1. Baskerville, R.: Investigating Information Systems with Action Research. Communications of the Association for Information Systems, 2 Article 19 (1999)
2. Bernard, S. A.: An Introduction to Enterprise Architecture - Linking Business and Technology. AuthorHouse, Indiana (2005) ISBN 1420880500
3. Briggs, R.O., Reinig, B.A., de Vreede, G.J.: Meeting Satisfaction for Technology-Supported Groups: An Empirical Validation of a Goal-Attainment Model. Small Group Research 37, 585–611 (2006), doi:10.1177/1046496406294320
4. Briggs, R.O., de Vreede, G.J., Nunamaker Jr., F.: Collaboration Engineering with ThinkLets to Pursue Sustained Success with Group Support Systems. Journal of Management Information Systems. 19, 31–64 (2003)
5. Briggs, R.O., de Vreede, G.J., Nunamaker Jr., J.F., Tobey, D.: Achieving Predictable, Repeatable, Patterns of Group Interaction with GSS. In: HICSS (2001)
6. Brynjolfsson, E., Saunders, A.: Wired for Innovation: How Information Technology is Reshaping the Economy. MIT Press, Cambridge (2010)
7. Gregor, S.: The Nature of Theory in Information Systems. MIS Quaterly 30(3), 611–642 (2006)
8. Henderson, J., Venkatraman, N.: Strategic alignment: Leveraging information technology for transforming organizations. IBM Syst. Journal 32(1), 4–16 (1993)
9. Hevner, A.R.: A Three Cycle View of Design Science Research. Scandinavian Journal of Information Systems 19(2), 87–92 (2007)
10. Hevner, A.R., March, S.T., Park, J., Ram, S.: Design Science in Information Systems Research. MIS Quarterly 28(1), 75–105 (2004)
11. Iivari, J.: A Paradigmatic Analysis of Information Systems as a Design Science. Scandinavian Journal of Information Systems 19(2), 39–64 (2007)
12. Janssen, M., Cresswell, A.: The Development of a Reference Architecture for Local Government. In: HICSS. IEEE Press, Los Alamitos (2005)
13. Kaisler, S.H., Armour, F., Valivullah, M.: Enterprise Architecting: Critical Problems. In: HICSS. IEEE Press, Los Alamitos (2005)
14. Kolfschoten, G.L., de Vreede, G.J.: Collaboration Engineering Approach for Designing Collaboration Processes. In: Haake, J.M., Ochoa, S.F., Cechich, A. (eds.) CRIWG 2007. LNCS, vol. 4715, pp. 95–110. Springer, Heidelberg (2007)
15. Lankhorst, M., et al.: Enterprise Architecture at Work: Modelling, Communication, and Analysis. Springer, Berlin (2005)

16. Lankhorst, M., van Drunen, H.: Enterprise Architecture Development and Modelling, http://www.via-nova-architectura.org
17. Mingers, J., Rosenhead, J.: Problem structuring methods in action. European Journal of Operational Research 152, 530–554 (2004)
18. Myers, M.D.: Qualitative Research in Information Systems. MIS Quarterly 21(2), 241–242 (1997)
19. Nakakawa, A.: Collaboration Engineering Approach to Enterprise Architecture Design Evaluation and Selection. In Proceedings of 15th CAiSE-DC (Doctoral Consortium) held in conjunction with CAiSE 2008, CEUR-WS, Montpellier, France vol. 343, pp. 85–94 (2008)
20. Nakakawa, A., van Bommel, P., Proper, H. A.: Quality Enhancement in Creating Enterprise Architecture: Relevance of Academic Models in Practice. In: E. Proper, F. Harmsen, and J.L.G. Dietz (Eds.): PRET 2009, LNBIP 28, pp. 109–133 (2009)
21. Nakakawa, A., van Bommel, P., Proper, H.A.: Requirements for Collaborative Decision Making in Enterprise Architecture. In: Proceedings of the 4th SIKS/BENAIS Conference on Enterprise Information Systems, The Netherlands, Nijmegen (2009)
22. Nakakawa, A., van Bommel, P., Proper, H.A.: Towards a Theory on Collaborative Decision Making in Enterprise Architecture. In: Winter, R., Zhao, J.L., Aier, S. (eds.) Global Perspectives on Design Science Research. LNCS, vol. 6105, pp. 538–541. Springer, Heidelberg (2010)
23. Nunamaker Jr., J.F., Briggs, R.O., Mittleman, D.D., Vogel, D.R., Balthazard, P.A.: Lessons from a dozen years of group support systems research: a discussion of lab and field findings. MIS 13(3), 163–207 (1996)
24. Op 't Land, M., Proper, H.A., Waage, M., Cloo, J., Steghuis, C.: Enterprise Architecture - Creating Value by Informed Governance. Springer, Berlin (2008) ISBN: 978-3-540-85231-5
25. Pervan, G., Lewis, L.F., Bajwa, D.S.: Adoption and use of electronic meeting systems in large Australian and New Zealand organizations. Group Decision and Negotiation 13, 403–414 (2004)
26. Proper, H.A., Hoppenbrouwers, S.J.B.A., Veldhuijzen van Zanten, G.E.: Communication of Enterprise Architectures. In: Lankhorst, M. (ed.) Enterprise Architecture at Work: Modeling, Communication and Analysis, pp. 67–82. Springer, Berlin (2005)
27. van der Raadt, B., Schouten, S., van Vliet, H.: Stakeholder Perspective of Enterprise Architecture. In: Morrison, R., Balasubramaniam, D., Falkner, K. (eds.) ECSA 2008. LNCS, vol. 5292, pp. 19–34. Springer, Heidelberg (2008)
28. Raiffa, H., Richardson, J., Metcalfe, D.: Negotiation Analysis - Science & Art of Collaborative Decision Making, Belknap Harvard, Cambridge, Massachusetts (2003)
29. Ross, J., Weill, P., Robertson, D.: Enterprise Architecture as Strategy: Creating a Foundation for Business Execution. Harvard Business School Press, Boston (2006)
30. Rouwette, E.A.J.A., Vennix, J.A.M., Felling, A.J.A.: On Evaluating the Performance of Problem Structuring Methods: An Attempt at Formulating a Conceptual Model. Group Decision and Negotiation 18(6), 567–587 (2007)
31. Schekkerman, J.: How to survive in the jungle of Enterprise Architecture Frameworks, Creating or Choosing an Enterprise Architecture Framework. Trafford Publishing, Canada (2004)
32. Simon, H.A.: The New Science of Management Decision. Harper and Row, New York (1960)

33. Spewak, S.H.: Enterprise Architecture Planning: Developing a Blue Print for Data, Applications, and Technology. John Wiley & Sons Inc., New York (1992)
34. Susman, G., Evered, R.: An Assessment of The Scientific Merits of Action Research. Administrative Science Quarterly 23(4), 582–603 (1978)
35. The Open Group Architecture Forum. TOGAF Version 9. Zaltbommel. Van Haren Publishing, The Netherlands (2009) ISBN: 978-90-8753-230-7
36. Zachman, J.: A framework for information systems architecture. IBM Systems Journal 26(3), 276–292 (1987)

Dealing with IT Risk in Nine Major Dutch Organizations

Theo Thiadens, Rien Hamers, Jacqueline van den Broek, Sander van Laar,
and Guido Coenders

Fontys University of Applied Sciences, Rachelsmolen 1, Postbus 347,
5600 AH Eindhoven, Netherlands
t.thiadens@fontys.nl

Abstract. In addition to architecture and IT portfolio management, IT risk is often mentioned as the third aspect that needs consideration when governing the
application of IT such that it optimally fits in with an organization's requirements (Dhillon et al [6]). This article investigates the degree of awareness with
respect to IT risks and the measures that are taken to reduce these risks in nine
large Dutch organizations. The study shows that IT users in these large organizations, faced with the question which risk they consider the most serious one,
all mention the lack of agility of their IT. Regarding the measures that are taken
for limiting risks, one may conclude that these large organizations often have
not organized IT risk management as a separate function that reports directly to
the senior management.

Keywords: IT risk, agility of IT, accuracy of data, availability of IT, access
to IT.

1 Introduction

In the year 2010, many organizations depend on the application of IT (Applegate [2]).
In some organizations, any supply of products and services without IT has even become impossible. Relying on the application of IT has become part and parcel in
present-day management. This management sets the priorities in an organization.
Does it demand maximum availability of IT? Does it demand 100% security when
using IT? And provided that it has better data at its disposal, would it be able to better
respond to the customer's wishes? Or does it perhaps wish IT to be more agile with
respect to its support by IT? The management of an organization weighs up the pros
and cons. In doing so, it has several choices. It may put an emphasis on availability of
IT, on protection when using IT, on the accuracy of data or on the increased agility
through application of IT. In this respect IT risk refers to the possibility of the occurrence of an unplanned event involving a failure or misuse of IT that threatens the
business objectives. The management of an organization governs IT. Transformation
of an organization often requires use of IT. So management decides which emphasis
is chosen when applying IT.

The objective of this article is to look at risk from an organizational level and to list
the measures as taken by organizations with regard to this risk. As a basis for this
study, we have started by listing the popular methods for looking at IT risk. From

F. Harmsen et al. (Eds.): PRET 2010, LNBIP 69, pp. 182–192, 2010.

these methods, we selected the method that includes the 4A and the 3CD model. Using this method, it was subsequently investigated which priorities the managers of nine large Dutch organizations put on the acceptance of risks in their use of IT. We also investigated what efforts their organizations make for meeting these priorities. The empirical part of this research took place in the first sixth months of 2009. The article starts by giving an overview of the theory. Next, it explains the set-up of the empirical part of the research, after which it gives an overview of the main results of this study (van den Broek et al [4]). Conclusions are drawn on this basis. The article concludes with a discussion. In this discussion, a link is made with the current Westerman et al [13] study.

2   IT Risk: The Investigated Perspectives

2.1   IT Risk: The Perspectives

In literature discussing the ways in which organizations may deal with the risks connected with the application of IT, one may come across various perspectives on the risks as run by organizations in the application of IT or when using IT supported information systems.

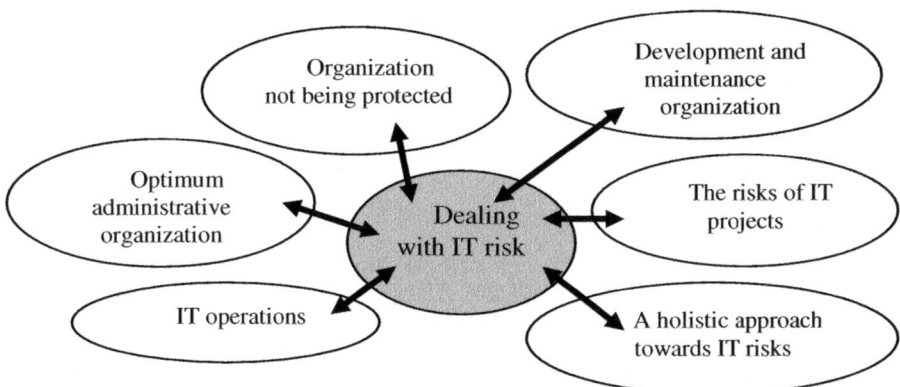

Fig. 1. Six perspectives when dealing with IT risk

Listing these perspectives reveals a distinction between (see figure 1, (Thiadens [12]):

1. The risk of the *organization not being protected* against the risks it may run regarding IT. In that case, one decides which risks an organization wants to cover itself against; at what moment the organization takes measures against the risks in question (e.g. ex ante or ex post) and what type of measures (e.g. logical, organizational or physical). The essence of this perspective is to safeguard protection against risks regarding confidentiality, reliability and continuity of the information provision, as well as the ICT required for this (Overbeek et al [9]; ISO 27000 [7]).

2. This risk is that the used information is not of the right quality. In this case, one focuses on making sure that the *optimum administrative organization* is achieved (Starreveld et al [11]; Romney [10]). The doctrine of the administrative organization states which measures an organization can take ex ante in order to make sure that the organization works with reliable information, observes the necessary confidentiality and that there are as few incursions on continuity of the information provision as possible. The doctrine also states how people check ex post, whether the rules with regard to competences in the field of information provision and the rules for ensuring reliable information have been observed.

3. The risk as run by an organization in development and maintenance of IT, as far as its IT organization is concerned. This involves examining the *organizational measures as taken by an organization in the field of development and maintenance of IT for limiting this risk.* (Meijer [8]). This leads to recommendations with regard to the manner in which various tasks in this field need to be organized and how to ensure that the desired organization is realized.

4. Attention for the *risk in IT operations* (de Wijs [14]). This is known as operational risk. De Wijs [14] investigated how organizations deal with the operational risks when their work is supported by IT. He formulated rules that are based on this and which lead to economically substantiated behaviour for dealing with this risk. He established that organizations do accept certain risks and in other cases take measures for minimizing risks as much as possible.

5. Attention for reducing *the risk that projects for developing and implementing IT provisions more or less fail* (Applegate et al [2]). Applegate et al state that participation in IT projects does involve risks. They state that these risks are subject to the size of the project, to the degree to which the requirements to the project are clear and whether the organization has the technical knowledge for completing the project at its disposal. Applegate et al specify which measures can be taken by an organization for concluding a project optimally based on a classification of projects.

6. Dealing with risk *from the perspective of the organization as a whole*. Examples of this approach are given by Bahli et al [3] and Westerman et al [13]. This approach is viewed as a more holistic approach.

Bahli et al [3] state that risk can be defined from two different perspectives. The first is the decision-theoretic view in which risk reflects the variance and gains associated with a particular alternative. The second is the behavioural perspective, which associates risk with the magnitude of a negative consequence of a decision. For their research Bahli et al [3] used the behavioural definition and looked at the negative consequences of business decisions, their likelihood and their associated impacts. To capture the components of risks they conceptualized risk as a set of triplets composed of scenarios: what can happen, the likelihood of a scenario and its consequences.

Westerman et al [13] developed a method to deal with IT risk, which is based on two cornerstones. These are: what does the management of an organization consider to be IT risk and what does this management do to limit this risk. The method distinguishes between the risks of availability, access, accuracy and agility of IT. And during its meetings, it enables the management of an organization to get clear from which perspective decisions on IT are made. One particular member of the board may for example operate from opportunities (agility), whilst another one puts an emphasis on control (access).

Of these six perspectives, this study has chosen the last one. This study looks at the risk that one runs at application of IT at the level of an organization and at the measurements with regard to this that the organization as a whole has taken. It follows that the methods as stated under perspectives 1 to 5 are of less because these only look at a specific risk of IT or at a certain aspect such as security. Deciding to view the IT risk from this organization perspective is also inspired by the fact that recent research teaches that the management in 85% of the companies think that their organization should reconsider their method for dealing with risk (Accenture [1]). Furthermore, Harvard Business Review's Daily Stat [5] states that these board members do not sufficiently include the risks they run in their decision-making process; that alignment between business strategy and risk lacks, that realization of this does involve members of the board often not having up-to-date and reliable data at their disposal for including risk assessment in their decision-making process.

As far as choosing between Bahli's and Westerman's method is concerned, we decided to go for the approach that can be tested by means of in-depth interviews with executives in an organization. In this study, IT risk have been viewed from the perspective of an organization and defined as (Westerman et al [13]): *"The possibility of an unplanned event as a result of the failing or incorrect use of ICT, which means that one or more of the organization's objectives are not achieved."*.

2.2   The Method as Proposed by Westerman et al.

The method of Westerman et al [13] is based on the definition of risks as experienced by managers in their daily practice and on an inventory of the measures as taken by the organization to deal with this risk. They define the IT risk as experienced by the managers at the user side of IT by means of the so-called 4A model. The measures for limiting the risk are researched and reproduced by means of the 3 Core Disciplines – further on called the 3CD- model.

The 4A model sums up the risks as run by an organization in four different areas, the so called 4 A's. These four A's are:

1. *Availability:* keep the systems (and their business processes) running, and recover from interruptions;
2. *Accessibility:* ensure appropriate access to data and systems so that only the right people have the access they need and the wrong people do not have access (the potential for misuse of sensitive information falls within this category).
3. *Accuracy:* provide correct, timely and complete information that meets the requirements of management, staff, customers, suppliers and regulators.
4. *Agility:* possessing the capability to change with managed cost and speed – for example, by acquiring a firm, completing a major process redesign or launching a new product/service. (IT consequences that constrict enterprise action fall within this category)

Organizations take measures for dealing with these risks. These measures can be divided into three types (3CD model). The study as performed by Westerman et al

[13] in 180 large companies shows these three types of measures. Each of these ensures transparency of the IT foundation, organizes optimal processes for risk management and ensures that employees are aware of the risks they have run and will run. Dealing with IT risks means that organizations look at the total risk and weigh up the pros and cons with respect to the measures to be taken (see figure 2). In general, this leads to:

a. a more transparent set-up of its *IT foundation*. This means a transparent architecture of products and services provided by a defined IT organization. This organization is no more complex than necessary.

b. the presence of an *organization for risk management*, which ensures the availability of an overview of the risks that are run at organizational level. This allows the management to invest sufficient time and means in risk management. In this process, risks are identified, given a specific priority and followed up.

c. and a *culture*, which ensures that everybody involved is sufficiently aware of the risks and takes these into account (risk awareness). In this culture, the risks that one may possibly run are discussed openly and non-threatening.

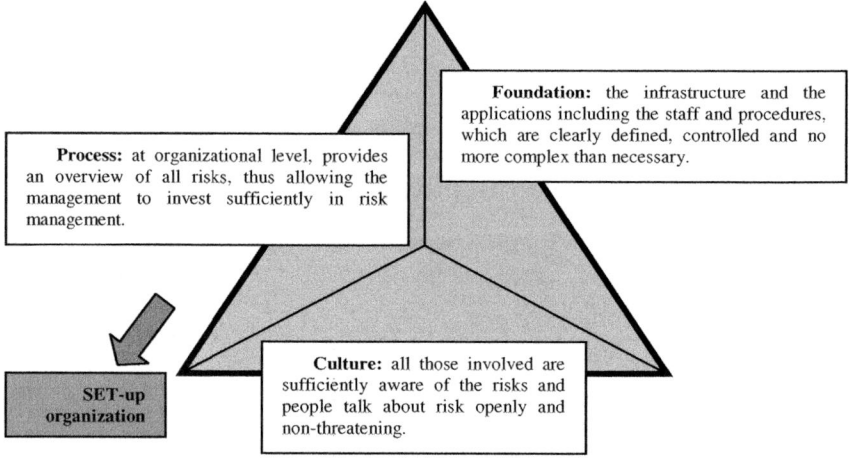

Fig. 2. Three ways to reduce IT risks

2.3 The Research Questions

Next, the following general research questions were defined for determining the risks that organizations are involved in when applying IT:

a. How does the management of the studied organization deal with the risks it faces through application of IT and what does it experience as the main risk?

b. How do the studied organizations flesh out the 3CD model? And with regard to this, do they deal with IT risk as being a risk that is organized for the organization together with all other risks at organizational level?

3 The Empirical Study

3.1 The Set-up of the Empirical Study

By means of in-depth interviews, the study investigated the measures as taken by the nine organizations for limiting the risks involved in the application of IT. In doing so, several choices were made. On the one hand, these choices concerned the size of the organization and on the other hand, they concerned the research method. Regarding the size of the organization, organizations with over 2500 employees were chosen. The reason behind this choice is the fact that in reviewing recent articles for international conferences, it turned out that in articles including surveys and where no distinction was made regarding the size of the organization, this often resulted in the management of the studied organizations stating that they regard security as the principal risk that they run at application of IT. The writers of this article experienced this differently in their every-day practice.

With regard to the research method, further choices were made. Firstly, it was decided to use in-depth interviews using predefined questions and open answer. The use of in-depth interviews fits in with the type of interviewee (higher management in the line and in IT) and with the fact that every study is part of the education at the Fontys University of Applied Sciences (Thiadens [12]). Students were trained to hold these interviews.

Besides, it was decided to interview line managers for research of the risks to be run (4A model) and to interview the ICT organization of companies about the measures that were taken (3CD). The figures 3 and 4 refer directly to the questions asked. In every organization, each of the interviewees was on average interviewed for an hour and a half. The interviews were recorded on tape, processed and the reports of the interviews were verified by the interviewees. These interviewees worked in the following organizations: an organisation working in the education sector, being the Fontys University of Applied Sciences; three organizations in the private sector being chip manufacturer NXP, Dutch mortgage advisor the Hypotheker and insurance company Achmea; a semi public organization being the Amphia hospital and four public sector organizations, being the Centraal Justitieel Incasso Bureau (CJIB) that translates as the Central Fine Collection Agency and is an implementing organisation of the Dutch Ministry of Justice; the Dienst Uitvoering Onderwijs, which is responsible for the execution of several acts and regulations, such as student grants; the Dutch Land Register and the Police Force.

The trial interviews for this study took place at the Fontys University ICT and at the Fontys University's ICT Services department. During the interviews, standard questionnaires were used. These questionnaires were sent to the interviewees in advance. The questions for the interviews were formulated by the supervision committee for this study. The Westerman et al [13] theory provided the basic questions for these questionnaires.

3.2 The Results at the User Side: The Customer's View on IT Risks

The results of the interviews with line managers of those parts of the organization that use IT are given in table 1. This figure provides an overview of the results of the interviews per organization. The figure shows that each interviewed line manager does know what IT means to their business. Each interviewed line manager realizes this and is able to state when the last IT breakdown took place and what damage this caused.

Table 1. Some aspects of IT risks, as seen from the IT users perspective

Subject: Organization:	Availability: How often breakdown in last 12 months?	Availability: Able to limit the damage?	Accessibility: Access in which manner?	Accuracy, timeliness & reliability of data: main data available?	Agility: how often do projects planned with IT exceed planned time or budget?
Fontys	Unplanned downtime 1.5 day breakdown of Sharepoint, apart from planned maintenance.	It is assumed that everything works. Otherwise manually.	Password and logon to entire environment.	No, schedule is by definition not good and insight into availability classrooms.	50% extension of budget and time in projects.
Amphia	Two to three times per month and then for a brief period.	Yes, breakdown is brief.	Logging in using name/password.	Patient data available; management info is lacking.	Estimations almost always too low.
Achmea	Not once.	Immediate start disaster recovery plan.	Name/password and everything is logged.	Yes, 7 times 24 hours.	Almost always.
NXP	Depends on the application and the location. Root cause analysis performed every time.	Yes, emergency fallback & continuity plan.	Profiles and application through boss.	Yes, for operational data.	Sometimes. Recently introduced project management method helps to monitor this.
Hypotheker	Once, for one hour.	Yes, there is emergency fallback. Max. 24 hours down.	Personal name and password.	Works on the basis of action list, but not completed actions not discovered, nor signaled.	Only fully and nation-wide installed, when application works stably.
IB-Group	Last year, the communication with the 6500 Dutch schools was disrupted.	There is an emergency fallback centre.	Via name/password.	Yes, around 98%. Also, not all data is provided on time by customers.	Often, reason: initial estimation project too low.
Kadaster	Once, but only a sub system. Impossible for everything to be down.	Emergency fallback limits this. Manual is impossible.	Profiles are worked on. Currently name/password.	By means of a workflow management system for meeting deadlines.	Often, even though we are using the project management method Prince-2.
Police Force	Rarely not available, paper backup in case of.	Processes are of lower level.	Via profiles.	Yes, for operational data. Management information not always there.	40-50% of all projects.
CJIB	Not once.	Normally, via the ITIL processes, otherwise business continuity plan.	Via name and password.	Yes, systems with a 98.9% availability.	60-70% of all projects.

From the answers of these line managers, one may conclude that the availability of IT is really not an issue in the investigated organizations. It is a different story when they are interviewed about the accessibility of IT. The use of profiles that clearly state what people in a certain capacity are allowed to do is certainly not generally accepted. It also proved that not all organizations are in the habit of fully logging all operations on their IT systems.

Looking at the quality of the information as provided, it turns out that the investigated organizations can improve their management information. Either it is not available or it is not sufficiently accurate or incomplete. Finally, it is established that the investigated organizations experience difficulties in delivering new applications on time and within budget. Supplementations of often more than 50% in both money and time are either explained away by poor estimations at the start of a project or by the fact that one does not enter into production until an application is fully stable.

After establishing the fact in the field of the 4A's, the interviewees were asked about their preference regarding improvement, when speaking of these 4A's. In this case, the interviewed managers in seven of the nine questioned organizations appeared to prefer agility. Only the deputy director of the Fontys University of Applied Sciences remarked that as far as he is concerned, the IT systems simply have to be available and that this is his prime concern. Furthermore, a number of managers remarked that, with regard to the accessibility of IT, the situation should remain liveable. In a hospital for example, one cannot oblige a doctor to log off every time, when logging on takes several minutes. The use of ID cards, iris scans and fingerprints could improve the ease of use as far as IT security is concerned.

3.3 Measures for Reducing IT Risks

The management in the IT departments of the investigated organizations was interviewed about the measures as taken to mitigate IT risk. These results are given in table 2. This figure provides an overall view of measures as taken by organizations for dealing with risks. In this case, measures are taken for arriving at an IT foundation that is as transparent as possible. Larger organizations, such as the ones taking part in this study, do standardize as far as their use of IT is concerned. Sometimes, these organizations do still have some legacy applications but these are gradually taken out of production (e.g. the Police Force). Furthermore, the organizations state that they do work with architectures, in which all the agreements for the set-up of and the objects as used in the IT foundation of the organization are defined. With regard to this, it must be noted that working with architecture apparently takes precedence over definition of a perspicuously written IT policy.

A majority of the interviewed organizations has a continuity plan and this plan is reviewed and tested periodically. Besides, everyone has data on the availability of IT at one's disposal.

When the IT managers were asked about their appreciation of IT, their answers strongly differed. Some organizations state that the operations part of the IT organization is held in high regard but that the IT organization does not always receive recognition for development and maintenance of applications.

The measures for organizing risk management as an independent discipline are often still in their infancy. The only exception being the insurance company where an organization for risk management is embedded and which reports to the top management of the company. Thinking about IT risk is often part of the task of a security department in an IT organization (NXP and Amphia hospital). Sometimes it only comes up for discussion when designing IT projects such as happens at the CJIB, the IB group and the Hypotheker. Regarding the methods as used for assessing IT risk, a diversity of approaches emerges.

And finally, the awareness regarding IT risks. This is present in each of the studied organizations. One is able to discuss risk openly. However, there is limited systematic and formal exchange of experiences in this field.

Table 2. The measures as taken by an organization

Subject: Organization:	IT measures: Standardization strategy for IT?	IT measures: Clear IT strategy for providing guidance?	IT measures: Continuity plan & review policy?	IT measures: Availability data:	IT measures: Does a customer organization understand IT?	Risk processes: How is risk management organized?	Risk processes : Method for assessment?	Awareness: Awareness that intellectual property and knowledge are of vital importance?
Fontys	Standard is the infrastructure and the conditions for using it. Applications not.	No, there are technical frameworks.	No, will be available after 2010.	Yes, 1 x per month reporting.	Operations is in high regard, the projects has a worse reputation.	No separate department or risk officer.	No	No but availability is considered.
Amphia	Standard is the infrastructure. Applications less.	Yes and procurement applies it.	No, has been planned for.	Yes but not widely distributed.	Operations of IT is in high regard.	Through IT steering committee.	Spark/sprint.	Not yet really.
Achmea	The technical infrastructure is standard.	Yes	Yes and tested 1x per year.	Yes, 99.8%. Report/wk.	Remains an issue.	Yes, team of risk officers.	DICE for projects.	Yes, there are guidelines.
NXP	Yes, but less in the the production environment.	Yes	Yes and 1x per year review.	Yes, standard in contracts.	There is a linking pin per department.	Part of IT organization.	ISF,Firm.	Yes built-in in project, otherwise awareness.
Hypotheker	Yes, the complete IT foundation is standardized.	No	Yes and update every 3 months.	Report every month.	A lot of communication takes place.	IT risk is issue especially around projects.	No	Yes, certainly within IT.
IB-Group	Infrastructure is standard and for application partly.	Works under architecture.	Yes and 2x yearly review.	Yes required for client.	Varies.	In the making.	Sessions with users, COSSO. Prince 2.	Yes but there is no intellectual property.
Kadaster	Yes, uniform operations environment.	Yes, architecture-based.	Yes, review 6x yearly, test 2x..	Yes, per month to client.	Difficult issue.	Standby & disaster recovery management.	For parts walk-through.	Yes, strongly focused in operations.
Police Force	Yes, for infrastructure and applications.	Yes, but no complete policy.	No, but there is a backup computer center.	Yes	Tricky issue.	Assessment by third party.	Via assessing and measuring.	Yes spearhead keeping knowledge up-to-date.
CJIB	Yes in principle, but there is still legacy.	Could be more explicit.	Yes 1x yearly test & review.	Yes, 98.9% availability.	A lot of tension here.	Through project leaders.	CRAMM with addition of a special security regulation based on ISO27000.	Yes but there are limits.

4 Conclusions

This article investigated the following questions:

a. How does the management of the studied organization deal with the risks it faces through application of IT and what does it experience as the main risk?
b. How do the studied organizations flesh out the 3CD model? And with regard to this, do they deal with IT risk as being a risk that is organized for the organization together with all other risks at organizational level?

Regarding question (a), it may be concluded that when looking at IT risks, the availability of IT does no longer present problems in the year 2009. However, dealing with data protection is a more important matter for concern. As far as the quality of the data is concerned, some organization could possibly gain the necessary as far as the quality of their management information is concerned. As regards agility, it becomes clear that no manager will implement new IT if this does not function properly. This is often the reason for overrunning budgets and deadlines.

Furthermore, the Fontys University of Applied Sciences remarks that every organization may put a different emphasis as far as IT risk is concerned. As a university of Applied Sciences, Fontys firstly demands availability of its IT but also sees that better data quality would benefit its work. However, most organizations clearly have a higher degree of agility of IT at the top of their priorities' wish list.

With regard to question (b), it may be concluded that with respect to the measures for limiting the risks involved in IT, it becomes clear that the studied organizations strive for working under architecture and do this to a high degree and that in doing so, they strongly standardize. Furthermore, it is obvious that the operations of IT is often considered as a general and technical support service. This service is better appreciated than the delivery of development and maintenance services. Getting the customer to understand the value of the development and maintenance services does seem to present some challenges. Only the insurance company has created a separate organization for risk management. Where necessary, the companies do seem to be aware of the risks they run when applying IT.

The answers to the two research questions lead to a final conclusion. This conclusion may be that, as IT becomes more important, the attention of IT customers in these larger organizations is more focused on the agility of the application of IT. These large organizations do take measures, such as a larger degree of standardization of IT and do work using architecture. With regard to set-up of a separate organization for risk management, they are only at the start and they will have to make an effort in order to keep alive the awareness of risks as involved in the application of IT.

5 Discussion

On closer examination of the results of the study, it may be concluded that in the interviews regarding the measures for reducing the risk as involved in IT, only IT managers were interviewed. There was no alternative for this because of the way these organizations, where ensuring the reduction of IT risk is still predominantly an IT affair, have currently organized how they deal with IT risk. Furthermore, my remark, that the conclusions of the study are based on a small number of organizations.

Acknowledgement

The research for this article was carried out within the framework of a study into IT risk by the lectureship ICT governance of the Fontys University of Applied Sciences. The authors would like to thank all colleagues in the knowledge circle, the students and the organizations that participated in this study.

References

1. Accenture: Managing Risk for High Performance in Extraordinary Times: Report on the Accenture 2009 Global Risk Management Study (2009),
   http://www.accenture.com

2. Applegate, L.M., et al.: Corporate information strategy and management, text and cases, 8th edn. Irwin/McGrawhill, New York (2008)
3. Bahli, B., et al.: An Assessment of Information technology outsourcing risk. In: Proceedings of ICIS 2001 (2001)
4. van den Broek, J., et al.: Omgaan met IT risico, Eindhoven. Fontys IT governance serie (2009)
5. Daily stat: IT risk and company management. Harvard Business Publishing (July 16, 2009)
6. Dhillon, G., Coos, D., Paton, D.: Chapter 11: Strategic IT/IS Leadership and IT governance. In: Grant, K., Hackney, R., Edgar, D. (eds.) Strategic Information systems management. Cenacge learning, Andover (2010)
7. ISO: the 27000 series of norms (2008), http://www.27000.org/index.htm
8. Meijer, J.: Risico management binnen een ontwikkel- en beheerafdeling. Scriptie Open Universiteit, Heerlen (2007)
9. Overbeek, P., Lindgren de Roos, E.R., Spruit, M.: Informatiebeveiliging onder controle, 4th edn. Prentice Hall/Pearson education, Amsterdam (2008)
10. Romney, M.B., Steinbart, P.J.: Accounting Information Systems. Pearson Education, Amsterdam (2008)
11. Starreveld, van Leeuwen, van Nimwegen: Bestuurlijke Informatieverzorging, deel 1: Algemene grondslagen, 5th edn. Stenfert/Kroese, Groningen (2003)
12. Thiadens, T.J.G.: Method of research. Fontys university of applied sciences, Eindhoven (2010),
http://www.ict-management.com/eng/beheer/Fontys%20onderzoek.htm
13. Westerman, G., Hunter, R.: IT risk, turning business threats into competitive advantage. Harvard Business School press, Boston (2007)
14. de Wijs, C.: Information systems management in complex organizations. De Wijs, Voorburg (1995)

Lines in the Water

The Line of Reasoning in an Enterprise Engineering Case Study from the Public Sector

H.A. Erik Proper[1,2] and M. Op 't Land[3,4,5]

[1] Public Research Centre – Henri Tudor, Luxembourg
[2] Radboud University Nijmegen, Nijmegen, The Netherlands
[3] Capgemini, Utrecht, The Netherlands
[4] Technical University of Lisbon, Lisbon, Portugal
[5] Delft University of Technology, Delft, The Netherlands

Abstract. Present day enterprises face many challenges, including mergers, acquisitions, technological innovations and the quest for new business models. These developments pose several fundamental design challenges to enterprises. We take the perspective that the design of an enterprise essentially involves a rational process that is driven by creativity and communication. Being a rational process means that there should be some underlying line of reasoning in terms of verifiable assumptions about the environment, the requirements that should be met, trade-offs with regards to the alignment between different aspects of the enterprise, et cetera, while all being used to motivate scoping and design decisions.

The core driver for the research reported in this paper is the desire to better understand the line of reasoning as it is used in real-life enterprise engineering / architecture engagements. By documenting and studying the lines of reasoning followed in different cases, we aim to gain more insight into the actual reasoning followed in practical situations. This insight can then, for example, be contrasted to the line of reasoning as suggested by existing enterprise engineering / architecture methods, and more importantly be used to create more effective lines of reasoning in future cases.

The larger part of this paper is therefore dedicated to a discussion of a real-life case from the public sector, where we focus on the line of reasoning followed in this case. The case concerns a large transformation program in the context of *Rijkswaterstaat*, which is an agency of the Dutch Ministry of Transport, Public Works and Water Management. In discussing this case study, we will focus on the line of reasoning as it was actually followed by Rijkswaterstaat, while also briefly discussing some of the results that have been produced 'along the way'.

To be able to position / track the line of reasoning as it was followed in the case study, we also provide six possible *reasoning dimensions* along which we expect the line of reasoning to propagate. For each of these *reasoning dimensions* we will argue why it would be relevant to study its occurrence / use in real-life cases. When combined, these reasoning dimensions form a kind of a *reasoning map* for enterprise engineering / architecture. This map provides us with a basic a-priori understanding of the dimensions along which the line of reasoning followed in a specific case study may propagate. In discussing the Rijkswaterstaat case study we will indeed position the followed line of reasoning in relation to this *reasoning map*.

F. Harmsen et al. (Eds.): PRET 2010, LNBIP 69, pp. 193–216, 2010.

1   Introduction

Present day enterprises face many challenges, such as the recent economic turmoil, mergers, acquisitions, technological innovations, deregulation of international trade, privatisation of state owned companies and agencies, increased global competition and the quest for new business models. These changes are fuelled even more by the advances in eCommerce, Networked Business, Virtual Enterprises, Mashup Corporations, the availability of resources on a global scale, et cetera [1, 2, 3, 4]. Such factors all contribute towards an increasingly dynamic environment in which enterprises aim to thrive.

As a result, enterprises need to be agile to improve their chances of survival. In other words, they need the ability to quickly adapt themselves to changes in their environment, and seize opportunities as they avail themselves. Such agility has become a business requirement in many lines of business, from the defence industry (schedules for combat systems have shrunk from 8 to only 2 years) via the car industry (from concept to production, for a new model in a few months instead of 6 years) to the banking industry (time to market for a new product from 9–12 months to a few weeks).

These trends also trigger enterprises to re-structure themselves into specialised parts increasing the agility of the organisation as a whole [5, 6, 7]. Traditional fixed organisational structures are replaced by more dynamic *networked enterprises* [8, 9, 10, 11]. Such trends are certainly not limited to the private sector alone. Citizens and companies increasingly expect governments to operate more effectively and efficiently. This triggers governments to restructure the way they operate into more "customer focussed" agencies, while privatising executive agencies.

The above discussed developments pose fundamental design challenges to enterprises. For example, deliberate decisions have to be made on the division of tasks and responsibilities in networked enterprises, including topics such as business process outsourcing (BPO), and the use of shared-service centres (SSC) and cloud computing for (IT) services.

We take the perspective that at the core, the design of an enterprise (from its value propositions, via its business processes to its supporting IT) is a rational process which is driven by creativity and communication [12]. The American Engineers' Council for Professional Development [13] also refers to the duality between creativity and rationality by stating that engineering concerns "*the creative application of scientific principles to design or develop structures, machines, apparatus, or manufacturing processes, or works utilising them ...*". In other words, there should be some underlying line of reasoning in terms of verifiable assumptions about the environment, the requirements that should be met, trade-offs with regards to the alignment between different aspects of the enterprise, et cetera, while all being used to motivate scoping and design decisions. Some authors even compare the design of a system (such as an enterprise) to the *creation* of a mathematical proof [14].

In past real-world experiences [15, 16, 17, 18, 19, 20], we have also found that when (re)designing enterprises (and their IT support) it is highly beneficial to make a clear and fundamental distinction between core aspects, such as stakeholder motivations, desired functionalities, implementation independent construction, the actual implementation, system types, et cetera, but also to trace the line of reasoning that seems to flow between these aspects.

The line of reasoning that underpins the design of a (part of an) enterprise, may be constructed *a-priori* or *a-posteriori*. In the *a-priori* case, design decisions are constructed rationally in the sense that they are based purely on rational conjectures. In the *a-posteriori* case, design decisions are essentially made first and are then motivated or tested (in terms of their falsifiability). In sum, we think it is fair to state that when enterprises are (re)designed, the designers will construct an a-priori or a-posteriori *line of reasoning* to motivate the resulting design.

Our underlying *research* driver is the desire to better understand the line of reasoning as it is used in real world enterprise engineering / architecture engagements. Our underlying *practical* driver is the desire to further professionalise the field of enterprise engineering. By documenting and studying the lines of reasoning used in different cases, we aim to gain more insight into the actual reasoning followed in practical situations. This insight can then, for example, be contrasted to the line of reasoning as suggested by existing enterprise engineering / architecture methods, and more importantly be used to create more effective lines of reasoning in future cases.

This paper will therefore start (in Section 2) by discussing six possible dimensions that may be followed by the line of reasoning. For each of these *reasoning dimensions* we will argue why it would be relevant to study its occurrence / use in the real world. When combined, these reasoning dimensions essentially form a *reasoning map* for enterprise engineering. This map provides us with a basic a-priori understanding of the dimensions along which the line of reasoning followed in a specific case study may propagate. The map, however, is by no means intended as an a-priori limitation of the reasoning dimensions we will look for in future case-studies. As we conduct more case studies, we also expect the reasoning map to evolve.

The larger part of this paper is dedicated to the discussion of a case study from the public sector. The case study concerns a large transformation program in the context of *Rijkswaterstaat*, which is an agency of the Dutch Ministry of Transport, Public Works and Water Management. The *Rijkswaterstaat* agency (more details to follow in Section 3) is responsible for the construction, management, development and maintenance of the main infrastructure networks in the Netherlands. Needless to say that we can only touch upon some of the highlights of this case, since it involves a multi-year transformation program at the Rijkswaterstaat agency. In discussing this case study, we focus on the line of reasoning as it was actually followed in the case, most notably in terms of its so-called *DAShboard* (Section 4), while also briefly discussing some of the results that have been produced 'along the way' (Section 5).

2   The Reasoning Map

In this section we discuss the six *reasoning dimensions* in which we have an a-priori interest when investigating the line of reasoning that is followed in case studies. When combining these *reasoning dimensions*, they provide a *reasoning map* upon which the line of reasoning, as it is used in specific cases, can indeed be mapped.

Please note that the *reasoning map* should not be confused with an (attempt to create yet another) enterprise architecture / engineering framework. We also do not claim the dimensions as included in the reasoning map to be 'complete' in any way. They are purely intended as a starting point to 'make sense' out of the lines of reasoning followed in specific cases. We will, however, for each of the six *reasoning dimensions* motivate why we consider it to be relevant to study how the line of reasoning in specific cases propagates, while also arguing the potential added value of making the reasoning in this dimension explicit. This also implies that we will only focus on dimensions which indeed support the creation of a line of reasoning. Over time, and based on multiple case studies, we may draw the conclusion that certain dimensions are missing from existing architecture / engineering frameworks, or conversely, that certain reasoning dimensions are not used much in practice. The latter would suggest that these dimensions are less important or even superfluous.

2.1 Design Motivation

We regard an enterprise as a goal-oriented cooperative of people and means. This is in line with common definitions of organisation, e.g. *"organisations are (1) social entities, (2) directed towards a goal, (3) designed as systems of consciously structured and coordinated activities, and (4) connected with the external environment"* [21]. When indeed taking the view than an enterprise is a goal-oriented cooperative, we regard it as relevant to see if a goal-oriented *design motivation* dimension is present in the line of reasoning followed by real world cases.

In this reasoning dimension, we currently make a distinction between *motivation, requirements* and *design*. In the field of software engineering, this dimension comes mainly to the fore from the field of goal-oriented requirements engineering [22, 23, 24, 25]. The *motivation* is captured in terms of the goals of stakeholders, which provide the underpinning of the *requirements* that should be met by the *construction* of the system. We currently expect these motivations to involve (at least) four classes of goals:

1. What are the goals of stakeholders for owning / having the system?
2. What are the goals of stakeholders for transforming the system?
3. What are the goals of actors for playing a role in the system?
4. What are the goals of actors for using the system?

Note: with system we refer to the general systems theory's notion of system. We are not using it as a synonym to software application, as it seems to have become common place among IT people. Software applications indeed are systems, but enterprises and information systems are systems too.

The requirements can pertain to the services/functions offered by the system being designed, the qualities of these services with regards to their delivery (e.g. availability and security), qualities pertaining to their upkeep (e.g. costs and maintainability), as well as qualities pertaining to their change (e.g. flexibility and scalability).

In the field of enterprise architecture, one typically makes a distinction between architecture principles and the actual 'architectural design' [26, 27, 28, 29, 30, 31]. As discussed in [32], architecture principles are normative principles that limit the *design freedom* and provide a first translation of stakeholder's requirements towards a focussing of the design space. Therefore, normative principles take the form of *declarative statements* on essential properties of the system. This also implies the need for statements that provide more tangible guidance to the implementers, as well as allow for analysis of the design to assess whether (in particular qualitative) requirements are met. In other words, instructive statements which more tangibly express *how* the system is to be constructed, e.g. in terms of value exchanges, transactions, services, contracts, processes, components, objects, building blocks, et cetera. These instructive statements can then be used to represent the 'architectural design' of a system. In [32] it is proposed to refer to these statements as *instructions* since they tell designers specifically what to do and what not to do. This use of the word *instruction* also concurs with its definition from the dictionary [33]: *an outline or manual of (technical) procedure* Enterprises typically use models expressed in languages such as UML [34], BPMN [35], TOGAF's [30] content framework, ArchiMate [36], or the language suggested by the DEMO method [37], (as a base) to express such *instructions*.

In line with the commonly made distinction between architecture principles and the actual architectural design, we are therefore also interested to follow the line of reasoning between these two levels of design. In sum, the design motivation dimension therefore distinguishes: *motivation, requirements, normative design* and *instructive design*.

2.2 Implementation Abstraction

Traditionally, the field of information systems engineering [38, 39, 40] uses a distinction between an implementation free design of the information system, referred to as the *conceptual model* and one or two (increasingly) implementation dependent levels, usually referred to as the *logical* and *physical levels*. The Zachman framework [41] for *information systems* architecture (i.e. not for enterprise architecture) also reflects this distinction in terms of a conceptual level in terms of a business model from the perspective of the owner of the information system, a logical level in terms of an (information) systems model from the perspective of the designer, and a physical level in terms of technology model from the perspective of the builder.

In the field of software engineering a similar distinction has emerged in the context of MDA (Model-Driven Architecture) [42], where a distinction is made between a Computation-Independent Model (CIM) which essentially provides an (information technology) implementation independent view on a domain, Platform-Independent Model (PIM) describing the implementation in terms of behaviour and structure of applications regardless of the chosen (information technology) implementation platform, and a Platform-Specific Model (PSM) which contains all required information regarding a specific platform that developers may use to implement the executable code.

In the field of enterprise architecture and enterprise engineering we also see the suggestion to distinguish between an implementation independent level and an implementation dependent level. The Integrated Architecture Framework (IAF), as developed by Capgemini [43], distinguishes a conceptual, logical and physical level. In TOGAF [30] we see the *logical level* represented in terms of so-called *architecture building blocks* and the *physical level* as *solution building blocks*. The DEMO [37] methodology for enterprise engineering identifies an implementation independent level in terms of an *ontological model* of the enterprise, next to its *implementation model*.

From a research perspective we are interested to see if a distinction between an implementation independent level and an implementation dependent level is indeed present in the line of reasoning used in practice. At present we will not make a distinction between a logical and a physical implementation level, as the borderline between the two seems to be rather difficult to make. We will use the terms *essential model* and *implementation model* to refer to the implementation independent and implementation specific model respectively. When considering the two key meanings of the word *essence* as provided by the dictionary [33]:

1. the permanent as contrasted with the accidental element of being,
2. the individual, real, or ultimate nature of a thing especially as opposed to its existence.

we believe that *essential model* best captures the intention of an implementation independent model. Terminology such as conceptual (*of, relating to, or consisting of concepts* [33]) or ontological (*relating to or based upon being or existence* [33]) apply equally well to the implementation independent level as well as the implementation dependent level (which also exists, and can be described in terms of concepts).

Our scope is the design of enterprises. In other words, not just the design of information systems, and most certainly not just the design of computerised information systems. An enterprise is a socio-technical system. In other words, it primarily consists of human actors that are supported by different forms of technologies (including information technology). We take the stance that the role of technology in an enterprise is always supportive. More specifically, technology can never be (legally, morally, ethically, et cetera) responsible for its own actions. Whichever the level of technological support, human beings remain responsible. When an ATM at a bank 'thinks' it should not issue money to us, we might sometimes express our frustrations by 'vandalising' this piece of technology. However, human beings remain responsible for specifying the business rules used by the ATM's software to determine that it will not give us money. In that sense we should really be 'vandalising' them. For example, in DEMO [37] this is made explicit by stating that an enterprise is primarily being a social systems in which the core elements are social individuals, where the operating principle is the fact that the constituent social individuals enter into and comply with commitments regarding the products or services to be created or delivered.

When regarding an enterprise as consisting of human actors supported by technology, one does need a refined view on the notion of *implementation model*. The implementation of the *essential model* of an enterprise involves the "implementation" of responsibilities in terms of human actors and the implementation of technologies that support these human actors in their responsibilities. In that sense, the ATM of a bank supports cashiers in their responsibility of issue cash to clients. The implementation in terms of responsibilities for human actors is what we will refer to as the *social implementation* and the implementation in terms of the underlying technologies as the *technological implementation*. The *social implementation* focuses on the division of the essential tasks and responsibilities identified at the ontological level among human actors, while the *technological implementation* then identifies the technological means that can be used by these human actors to support them in their tasks and responsibilities.

We expect that identifying a *social implementation* level also invites designers of enterprises to carefully think about the impact of their design decisions on the well being of the human actors. In doing so, one would expect designers to also take properties such as work load, ethical burden, cognitive load, et cetera, into consideration in weighing between design alternatives. To illustrate this point, consider a small example in terms of a Pizza delivery service. At an ontological level (see e.g. [28]), the driving transaction of a Pizza delivery service is the ordering and delivering of a Pizza. In the essential model, this corresponds to a single transaction *complete Pizza purchase* involving an actor (role) *customer* and *Pizza order completer*. At the essential level, the costumer requests a Pizza, while the completer delivers the Pizza.

Now consider the following commonly used social implementation of this essential model. The taking of the Pizza order is done by some *order taker* functionary who answers phone calls. The Pizza is then baked by a functionary *baker*, and delivered by a functionary *deliverer*. This means that the essential transaction of *complete Pizza purchase* as it is performed by the single (essential) actor *Pizza order completer*, is actually implemented in terms of three functionary types. And indeed, in most practical situations this is highly defendable. It allows for an efficient use of time, technological means and furthermore enables an effective build up of specialised skills. The disadvantage of this implementation is, however, is the commitment to deliver (the right!) Pizza to the customer is done by another functionary than the functionary who actually delivers the Pizza. As a consequence, when the wrong Pizza is produced, or the Pizza is baked incorrectly, the Pizza deliverer is confronted with the angry customer while the deliverer can hardly influence these qualities. When this happens often, this is likely to lead to stress and general negative feelings with those who execute the functionary role of Pizza deliverer.

An alternative social implementation would be to only have a *Pizza order completer* functionary. As a consequence, the person who takes the order, would also be the one baking your Pizza, and then driving out to your house to personally deliver the Pizza. This is likely to be inefficient from a time and resource perspective. Nevertheless, in the case of failures, only the directly responsible person is confronted with the complaints / angry client. A hybrid implementation would, for example, be to indeed use the first division into multiple functionary types, but to ensure regular job rotation among the human beings who execute the different functionary types.

Even though we have theoretical and ethical reasons to expect / want to find a distinction between a social and a technological implementation, we expect to find little of such a distinction in practice. We base this scepticism mainly on the lack of such a distinction in existing architecture frameworks and engineering frameworks. In our view, most (if not all) of these frameworks take a technology-minded perspective on the implementation model. So, even if a distinction is made between an essential level and an implementation level, the latter is mainly focusing on the technology implementation. Nevertheless, we hope to find at least 'traces' of taking properties such as work load, ethical burden, cognitive load, et cetera, into consideration in weighing between design alternatives.

In sum, the implementation abstraction dimension therefore distinguishes: the *essential level*, the *social implementation level*, and the *technological implementation level*.

2.3   Construction Abstraction

In the engineering of systems in general, a well known distinction is the one between a *black box* and a *white box* perspective[1]. Typically, from a *black box* perspective, one regards a system (such as an IT system, an information system, or an enterprise) solely in terms of its input, output and transfer characteristics without any knowledge of its internal workings. In other words, its internal construction is "opaque" (black). The opposite of a *black box* perspective on a system, is the *white box* perspective where the inner construction of the system can indeed be observed.

For example, in the DEMO [37] methodology for enterprise engineering, this distinction is made explicit in terms of the *function* and *construction* perspectives. In the ArchiMate [36] standard, this distinction comes to the fore in terms of the *internal* and *external* perspectives.

The construction abstraction dimension, therefore distinguishes: the *black-box perspective* and the *white-box perspective*.

2.4   System Types

We are interested in knowing the system types that are being used in specific cases, as well as how these system types are linked. For example, the DEMO [37] methodology identifies a *B-organisation*, *I-organisation* and *D-organisation*, representing a system type focussing on business processes, information processing and data processing respectively. AchiMate [36] distinguishes a *business layer*, *application layer* and *technology layer*, representing system types focussing on business processes, computerised information processing, and the underlying IT infrastructure, respectively. Similarly, the Integrated Architecture Framework [43] (also used as a base in the Rijkswaterstaat case) distinguishes between business activities, information (processing) needed for the business, computerised information systems, and the technology infrastructure needed for these latter systems.

This reasoning dimension does not have a pre-defined set of values. There seems to be a general understanding in (IT focussed) approaches that there is a general *business system* level that uses *computerised information systems*, which on their turn depend on underlying *infrastructure systems*. However, since the field of enterprise architecture / engineering increasingly moves beyond the IT centric focus, we think it is not wise to a-priori fix the values in this dimension, and rather observe the values used in practice.

2.5   Design Evolution

Enterprises are hardly ever created from scratch. In other words, one has to deal with existing products, processes, information systems, et cetera. This also implies that when design a new step in the evolution of an enterprise, one cannot just look at the future 'version' of the enterprise in isolation. One has to consider the existing situation (and its past) to understand / rationalise some of the design decisions underlying the next steps in the evolution of the enterprise.

---

[1] See e.g. http://en.wikipedia.org/wiki/Black_box_(systems).

The TOGAF methodology [30] traditionally makes a distinction between a *baseline architecture* and a *target architecture*. Other sources may refer to the *ist* and *soll* situation. More recently, TOGAF also introduced the concept of *transition architecture* to allow for the fact that a transformation from the baseline situation towards the target situation, is likely to involve multiple steps (or plateaux).

Similarly to the system types, we currently do not provide a pre-defined set of values. We are also interested in observing the kinds of values used in real case studies. Generally, however, one would expect three types of values: (1) *initial state*, i.e. the design of the enterprise at the start of a (proposed) transformation, (2) *intermediary states*, several intermediary states in terms of e.g. plateaux and (3) *final state*, the design of the future enterprise as it is currently envisaged.

Within this dimension we are also interested to see if real-world cases indeed explicitly use knowledge of the existing situation to support design decisions on the final state. Even more, it will be interesting to see if the intermediary states will be motivated in terms of a gap analysis between the initiate state and final state, arguing how this gap will be bridged in a series of intermediary steps. Since the execution of enterprise transformations tend to take longer periods of time, it is also interesting to see if in the identification of intermediary states, one has taken into consideration that during the execution of an actual transformation the final state will change due to new requirements.

2.6   Design Horizon

This final dimension takes into account that in engineering / architecting an enterprise one can take a short term or a long term perspective. More specifically, it seems (a-priori) reasonable to distinguish between three levels:

Strategical design horizon – This is the level of the enterprise's strategy, including sub-strategies dealing with business aspects, human resourcing issues, IT, et cetera.

Tactical design horizon – This is the level at which the enterprise's strategy is made more concrete in terms of general requirements on, design principles for and high level designs of, (classes of) sub-systems within the enterprise, as well as the identification of programmes needed to execute a proposed transformations in terms of changes / creation of the identified sub-systems.

Operational design horizon – At this level, we are at the level of the design of specific sub-systems of the enterprise, projects filling in the enterprise transformation, et cetera.

The *tactical design horizon* might be referred to as the *architecture* level, while the *operational design horizon* might be called the traditional *design* level. However, to avoid confusion with our more general use of the word design, and to acknowledge the fact that in general the distinction between strategy, architecture and design is still open to debate, we refrain from using these words. At the same time, we do not claim that the strategical / tactical / operational distinction solves this. However, these terms do enable a more neutral observation of what happened in a specific case, separate from the discussion if the way a specific case used the term architecture is indeed correct from a specific definition of architecture.

3   Rijkswaterstaat and the Berthing-Place Domain

The case study reported on in this paper, was conducted at the agency Rijkswaterstaat (RWS), which is the Directorate-General for Public Works and Water Management in the Netherlands. Under the command of a departmental Minister and State Secretary, RWS is responsible for the construction, management, development and maintenance of some of the main infrastructure networks in the Netherlands, namely the networks for the transportation of water, road traffic and water traffic. Since a significant part of the Dutch economy is directly dependent on the logistics sector, including transport of goods from / to the Rotterdam and Amsterdam ports, as well as from/to the Schiphol airport, RWS has a very important role to play in the Netherlands.

In addition to the road and railway[2] networks, an important part of the logistical infrastructure in the Netherlands are its waterways. Several rivers run through the Netherlands, allowing goods to be transported by ships from / to Belgium, Germany and beyond. These rivers are also connected by several major canals. To regulate both the water flow and shipping traffic, several locks have been put in place. RWS is responsible for the maintenance and management of these waterways and associated infrastructure. A further responsibility of this executive agency is the management of the network of dikes, dams, and other means needed to keep the Netherlands from flooding. Needless to say that to a country which is positioned largely below sea level this is a task of some importance, especially when this is combined with one of the busiest network of waterways in the world.

The case which we focus on in this paper is concerned with a specific aspect of *Shipping Traffic Management* (SVM; In Dutch: ScheepvaartVerkeersManagement), namely the use of berthing places. A berth place is an area where a ship can dock; essentially a "parking spot" for ships. Some of these berths are used as holding areas at busy locks or bridges, while others are used for ships and their crews to stay overnight, et cetera. Not all berthing places are owned by the central government. A large number of them are owned by (e.g. municipal or provincial) port administrations, companies with their own ports and associated berthing places, et cetera.

Having good and sufficient berthing places is essential for efficiency and safety on the Dutch water ways. E.g. skippers not only need to take rest, but they are also lawfully obliged to do so. RWS is required to facilitate this. Failing to do so may lead to delays on the waterways and may even increase the risk of accidents. A survey held by RWS before the start of this case-study revealed that the users of berthing places were not satisfied with the amount, quality and location of berthing places. At the same time, the financial aspects should not be underestimated. Berthing places are quite costly to create and to maintain. Therefore, 'simply' creating more berthing places is also not the answer. In other words, proper management is needed, carefully weighing the needs in terms of efficiency and safety of shipping traffic, the interests of the skippers, and the financial aspects of creating and maintaining berthing places.

To better support the SVM activities, including the regulation of the use of berthing places, RWS initiated a long term transformation program involving the creating (and realisation) of a *Domain Architecture SVM* (DAS; In Dutch: DomeinArchitectuur

---

[2] The management of the Dutch railway network is not part of RWS's responsibilities.

Scheepvaartverkeersmanagement). The design of an improved management system with associated procedures and IT support, is part of this transformation program.

4 The Rijkswaterstaat *DAShboard*

Large organisations such as Rijkswaterstaat (RWS) continually change their products and services, quite often in close cooperation with other value-chain partners. Such changes in value propositions are likely to have a deep impact on the structures and operations of the organisation as well as its supporting IT. To timely, coherently and consistently govern the transformation processes required to implement such large changes, RWS applies two main instruments: (1) a standardised process for Integrated Governance and (2) a reasoning framework, called the *DAShboard* [44]. The first instrument prescribes RWS's standard approach to move from an initial idea via an architectural exploration and analysis of alternatives, to annual change plans. These latter plans including the way in which stakeholders should be involved and decisions should be taken. In the remainder of this section we will concentrate on the second instrument, the DAShboard. First we will introduce its roots and underlying concepts, relating it to some of Section 2's *reasoning dimensions*, then we will present its contents.

The DAShboard is the standard reasoning framework of RWS, as applied by the Domain Architecture *S*VM; hence the nickname *DAS*hboard. It builds primarily on the Integrated Architecture Framework (IAF) [43], from which it borrows its two main axes: Aspect Areas and Abstraction Levels, as well as the notion of the Artefacts. Compared with IAF three changes have been made: (1) a Transformation Level is added, (2) the Aspect Area "Information" is split into "Information delivery" and "Data Management" and (3) instead of listing Artefacts the DAShboards sums up a number of key *questions*. The Artefacts as defined in IAF can certainly be used to provide an answer to these questions, where the listed questions can provide a clearer focus enabling a more conscious selection of those Artefacts that are the most effective for answering these questions. Table 1 shows the DAShboard, made specific for the "Berthing places" case.

The DAShboard discerns five Abstraction Levels, which allow problems to be split into separate aspects, enabling a stepwise solution:

- *Contextual level*, answering the "Why" question, such as drivers, objectives, principles and scope;
- *Conceptual level*, answering the "What" question, what are the requirements , what services should the solution deliver;
- *Logical level*, answering the "How" question with an "ideal" solution;
- *Physical level*, answering the "With what" question with physical means: people, organisations,
- *Transformation level*, answering the "When" question by providing a transformation path from AS IS to TO BE, and its underpinning by a Business Case.

Compared with Section 2's *reasoning dimensions*, the Abstraction Levels used by RWS correspond to multiple dimensions at the same time. More specifically:

- The *contextual*, *conceptual*, *logical* and *physical* levels together cover the *design motivation* dimension.

Table 1. DAShboard, with questions for the case "Berthing places Management (BpM)"

Contextual				
What are the goals of BpM? Who are the internal and external stakeholders for BpM and what are their interests and requirements? What types of Berthing places can be discerned, and how are they used?	Which laws and regulations are applicable for, or influencing, BpM? What are the main changes in the environment of BpM? What policies exist concerning the Business/Information, Applications and Infrastructure in the area of BpM?		Which principles and standards are applicable to BpM? Which running programmes / projects are influencing BpM?	

Business	*Information*		*Applications*	*Technology Infrastructure*
	Information supply	*Data management*		
Conceptual				
What business services does BpM supply and use? What is the required quality of these business services? What business actors deliver these services, and which business actors and need these services?	What information is used by the business actors? What is the required quality of the information services? What information actors deliver these services?	What data are used by the information actors? What business actors do create the original facts? What is the required quality of the data services? What data actors deliver these data services?	What application services support the business, information and data actors? What is the required quality of these application services? What application component deliver these application services?	What infrastructural services do support the business, information and data actors and the application components? What is the required quality of these infrastructural services? What infrastructural component deliver these infrastructural services?
Logical				
Which business objects are observed or changed when delivering the business services? How do the processes of the business actors operate?	How are the information products composed by data objects? How do the processes of the information actors operate?	How are the data objects composed, and which states of the business objects do they concern? In which way do the processes of the data actors operate?	How are the application interfaces structured? In which way do the application components operate, what are their mutual interactions, and what is the interaction with human actors?	How are the infrastructural interfaces structured? In which way do the infrastructural components operate, what are their mutual interactions, and what are the interaction with human actors?
Physical				
With what people and means are the business actors implemented, and on what locations? What are the operational costs of this implementation?	With what people and means are the information actors implemented, and on what locations? What are the operational costs of this implementation?	With what people and means are the data actors implemented, and on what locations? What are the operational costs of this implementation?	With what software products have the application components been implemented, and on what locations? What are the operational costs of this implementation?	With what infrastructural products have the infrastructural components been implemented, and on what locations? What are the operational costs of this implementation?
Transformational				
What are the differences between AS IS and TO BE for the business?	What are the differences between AS IS and TO BE for information supply?	What are the differences between AS IS and TO BE for data management?	What are the differences between AS IS and TO BE for the applications?	What are the differences between AS IS and TO BE for the infrastructure?
What is the change plan for the transformation of Business, Information Supply, Data Management, Applications and Infrastructure?				

The *contextual level* provides the *motivation*, while also more explicitly adding the notion of scope. The *conceptual level* corresponds to the (functional) *requirements*. The *logical* and *physical* levels together cover the *design* where no explicit distinction is made between a *normative* and an *instructive* design.

− The distinction between the *conceptual* and *logical* levels also corresponds to the distinction between the *black box* and *white box* perspective, as described in the *construction abstraction* dimension.
− The combination of the *conceptual* and *logical* levels, and the *physical* level roughly corresponds to the dimension of *implementation abstraction*, where the *conceptual* and *logical* level provide an implementation independent (i.e. the *essential level*) design (covering both a black-box and white-box perspective), while the *physical* level represents the *implementation level*. No specific distinction is made between a social and a technical implementation.
− *Transformation* contains the dimension *Design evolution*, adding to that the notion of the Business Case.

These findings are summarised in Table 2.

Table 2. The DAShboard's abstraction levels and the reasoning dimensions

DAShboard	Reasoning dimensions			
Abstraction level	Design motivation	Implementation abstraction	Construction abstraction	Design evolution
Contextual	Motivation	All	All	
Conceptual	Requirements	All	Black-box	
Logical	Design	Essential	White-box	
Physical	Design	Implementation	White-box	
Transformational				All

The second dimension in the DAShboard concerns the Aspect Areas —which corresponds to the reasoning dimension *System types*:

− *Business* deals with the creation of new facts in reality, or observing the state of reality; e.g. building a bridge (material new fact), closing a deal (immaterial new fact) or monitoring / judging the state of the road;
− *Information* deals with the creation of information needed by the business for their situational awareness, derived from original or derived facts, including the data management needed for that; e.g. creating information on built bridges per period and region, derived from original facts for each built bridge, on which data have been managed. RWS has split this Aspect Area in two sub-Aspect Areas, namely *Information delivery* and Data management;
− *Applications* provide automated information systems, including their mutual and human interfaces, to support or execute part of Business, Information delivery and Data Management;
− *Technology Infrastructure* provides the ICT infrastructure — such as computing systems, network technology and database management systems — to make the applications work.

In terms of the *design horizon* reasoning dimension, the DAShboard could be used at any of the levels identified. In the SVM case, it was mainly used at the *tactical* and *operational* levels.

The DAShboard provides a systematic analysis of business processes, from their contextual aspects, via their business aspects to their implementation using people and technological means, covering both the present and future states. The general flow of reasoning followed by RWS when using the DAShboard [45], is as follows:

1. One starts by considering the context (i.e. the contextual level). What is the strategy. What are the goals of the organisation? What is the desired direction of the transformation?
2. One then continues with the business aspects. What are the business services offered, and what are the processes needed to deliver these?
3. Then one continues with the identification of the information needed to support these processes. What information is needed? What data needs to be gathered and stored to provide this information?
4. The next step is to asses to what extend the data processing and information supply is / can / should be computerised.
5. Finally, one assess which organisation units and people are responsible for the execution, management, and maintenance of the processes and their IT support, as well as the costs involved.

Answering the questions listed in the DAShboard, also leads to a multiple ambition levels for the transformation of the organisation. In other words, a series of future scenarios with increasing ambitions. DAS also enables a cost / benefits analysis of all aspects of the organisation. This provides RWS with the ability to study the impact of design decisions beforehand, and use as a means to reduce risks when actually transforming the organisation.

RWS is using the DAShboard as a reasoning framework to systematically detect the impact of intended changes. In earlier versions, the cells of the framework were filled with Artefacts, such as "actor model", "use cases", "object model", "business function model". The names of those Artefacts were not clear for managerial purposes, which raised questions such as *"what do I need use cases for?"*. Therefore the managerial questions itself were projected in the framework and made specific for the project at hand, enabling conscious choices about which questions to answer and which questions to let go – at least at this stage of decision making. As an example, Table 1 shows the DAShboard, made specific for the case "Berthing places". Generally speaking, "jumping squares" or "moving one diagonal step" in the DAShboard is discouraged; it means "jumping to solutions", which threatens the traceability of the work.

5   Applying the Framework to the Rijkswaterstaat Case

In this section we will discuss *some* of the results that were produced when using the DAShboard in the SVM case of Rijkswaterstaat. We have split the discussion of the results along the DAShboard's abstraction levels, also following the process used by Rijkswaterstaat.

5.1 Contextual

As discussed in [46], two core problems triggered the transformation programme at RWS:

- Berthing place users are unsatisfied about the existing situation (quantity and locations).
- RWS experiences capacity problems on their berthing place due to increasing recreational use, as well as ships avoiding paid alternatives.

Influenceable (by RWS) root causes that have been discerned are:

- Objectification of causes for dissatisfaction and necessity for additional capacity.
- No integrated national policy for management of berthing places (involving RWS and partners)

Berthing places need to be requested and allocated well in time. One emerging task in the SVM-responsibility, therefore, is the use and allocation of berthing places. An important question was also what level / type of management for berthing places is needed. Ideally, this could be solved by introducing a broker between organisations offering berthing places and ship owners needing berthing places. To this end, Rijkswaterstaat considered several solutions for the apparent scarcity of berthing places. One of the solutions could e.g. include a *Shared Service Centre* to accommodate the brokering and allocation tasks. This however would be a rather costly solution.

The original goal for applying the DAShboard for the management of berthing places was therefore:

> *Formulate several solution alternatives for the improvement of the management of berthing places, by Rijkswaterstaat in collaboration with other organisations, and clarify the consequences on the primary processes, information provisioning, data provisioning and IT support.*

A first result in this case study was the context diagram as shown in Figure 1 and the goal tree shown in Figure 2 These diagrams provide an answer to the DAShboard question *What are the goals of BpM? Who are the internal and external stakeholders for BpM and what are their interests and requirements? What types of Berthing places can be discerned, and how are they used?* The context diagram in Figure 1 provides an overview of the key stakeholders of shipping traffic management, while the goal tree included in Figure 2 zooms in on the goals of Rijkswaterstaat pertaining to safe and efficient shipping traffic, and the role of berthing places.

Based on the goals of stakeholders, and the possible impact of different designs of the enterprise, several architecture principles were formulated that address concerns raised by these goals:

- Rijkswaterstaat orchestrates the information flow in the traffic management chain.
- Data is acquired, stored and managed at one location only.
- Information systems facilitate coworkers in their roles and responsibilities.
- The presence of a berthing place is not allowed to decrease efficiency and safety of shipping traffic.

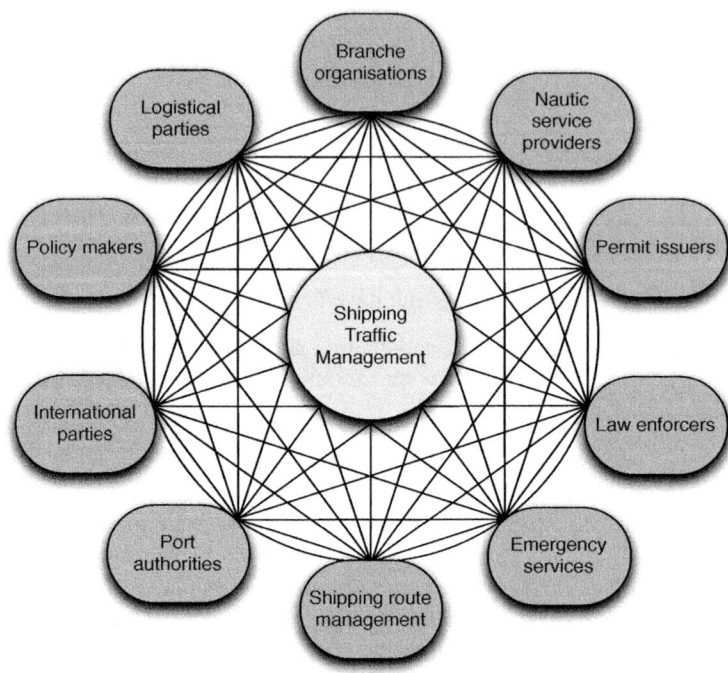

Fig. 1. Context of Shipping Traffic Management

In addition, since the Rijkswaterstaat agency is formally part of the Dutch government, all architecture principles included in the NORA, the Netherlands' Government Reference Architecture [50], apply as well.

5.2 Conceptual

The *function* perspective on the *Shipping Traffic Management* domain, i.e. the *black box* perspective, is provided in Figure 3. This diagram basically is a mind map which lists the functions provided by *Shipping Traffic Management* to its clients, which also shows the satisfaction of stakeholders in the current situation. For example, 40% of the captains of freight ships is satisfied with the number of berthing places available at locks.

5.3 Logical

The core of the logical level (the business aspect) is shown in Figure 4. The depicted construction model corresponds to the *white-box* perspective on core transactions of this domain. The notation used in Figure 4 is the notation for construction models as suggested in the DEMO method from [37]. Core to these models are the actors and the transactions in which they are involved. For example, the transaction labelled T01 leads to the result *berthing place reserved*. It is initiated by actor CA01, the *Berthing place user*, while the request will be met by actor A21, the *Berthing place reservation maker*.

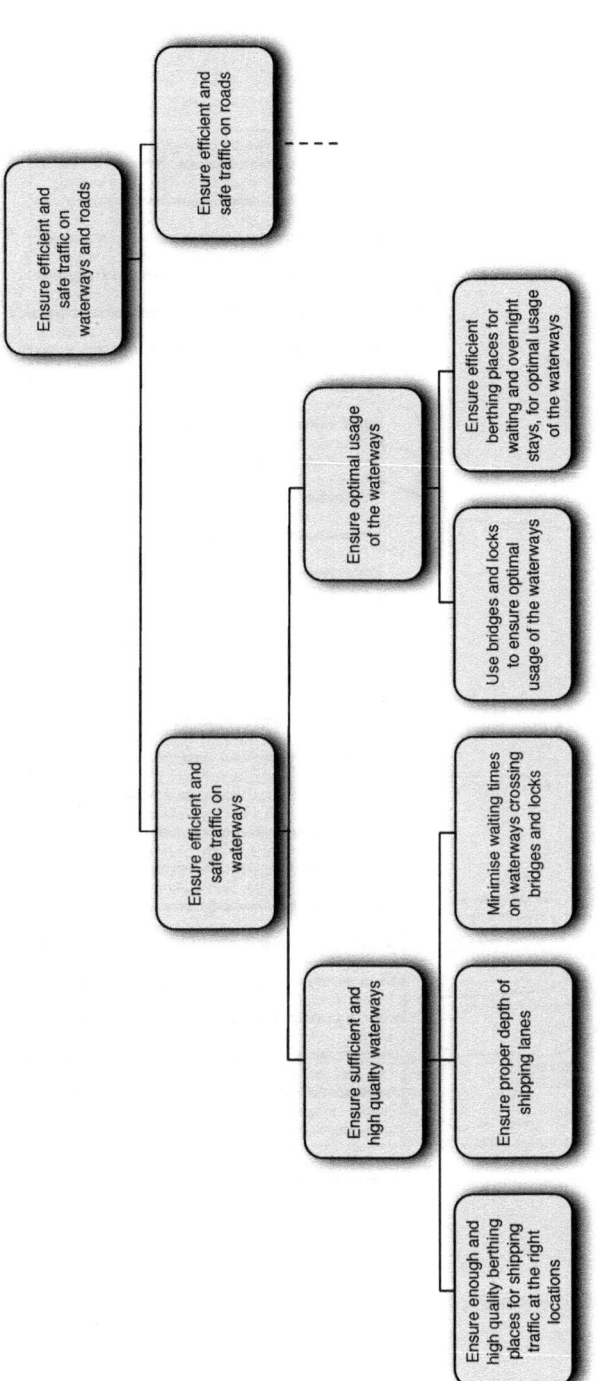

Fig. 2. Goal tree for (part of) the Rijkswaterstaat case

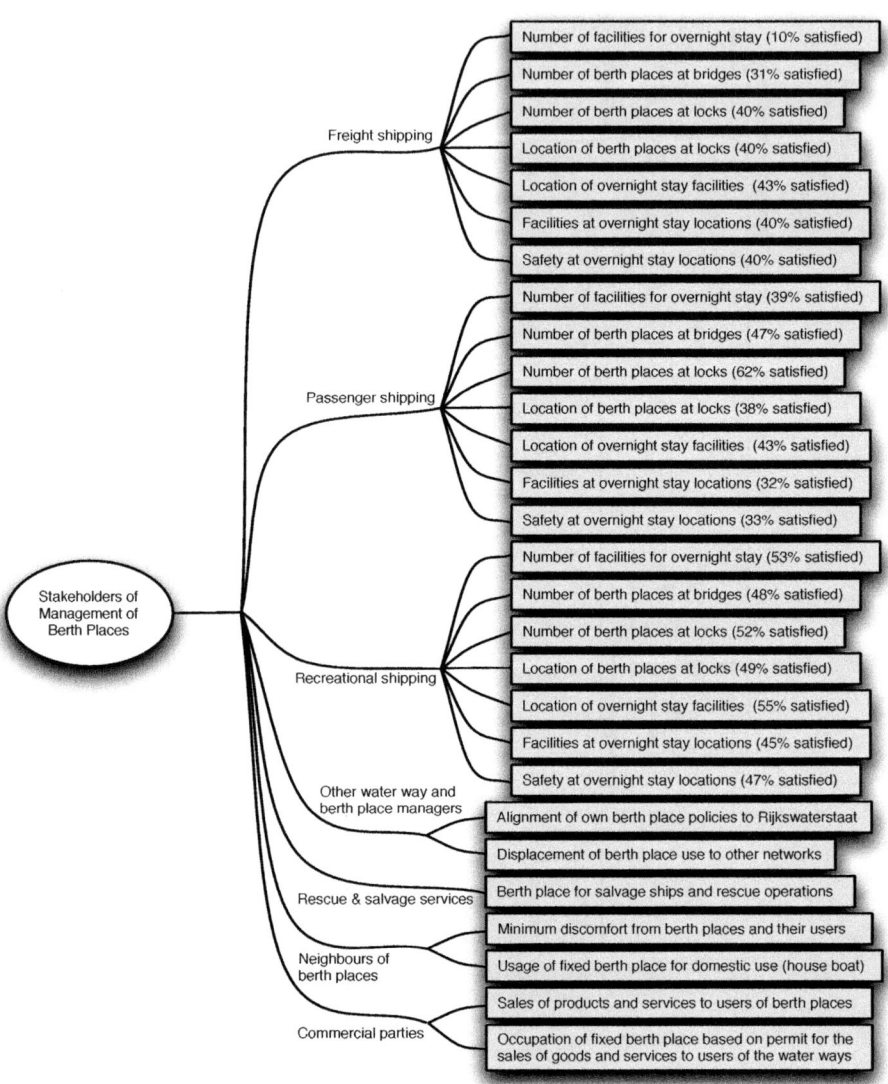

Fig. 3. Function design of the Rijkswaterstaat case

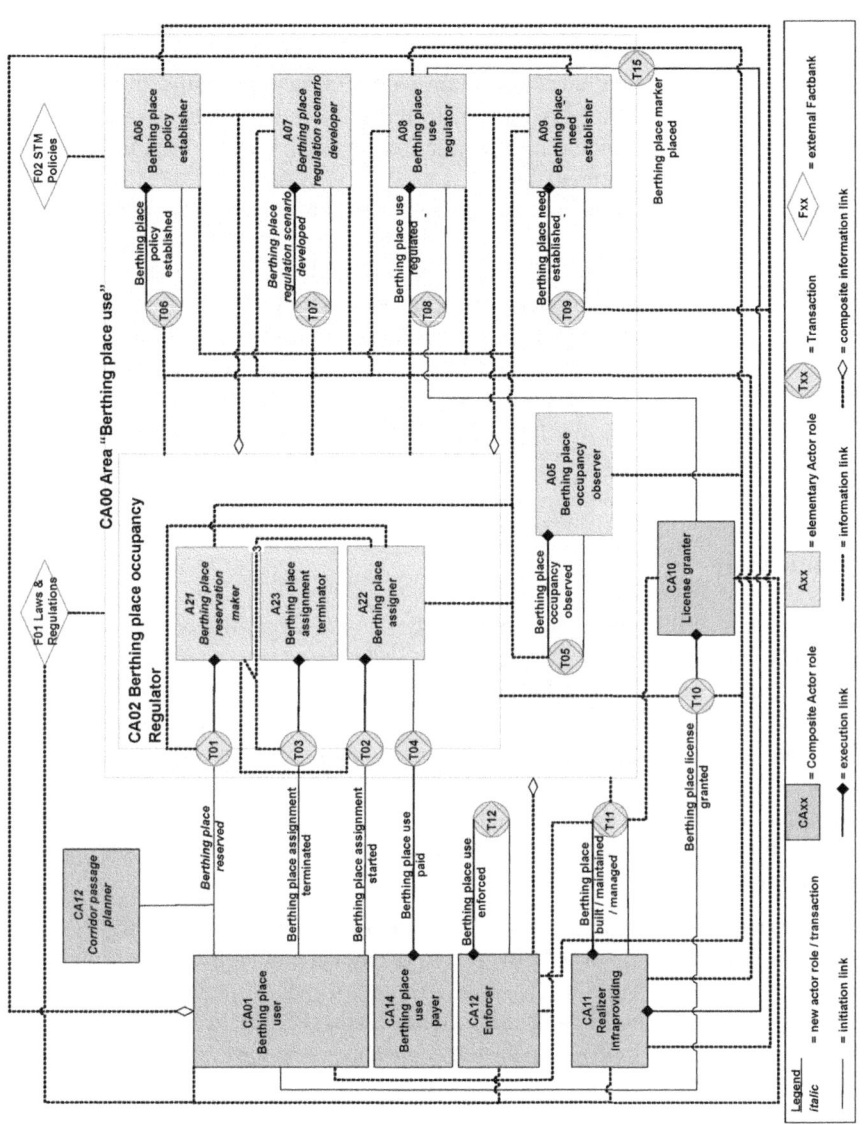

Fig. 4. Example logical level model from the Rijkswaterstaat case

5.4 Physical

In the Rijkswaterstaat case, several applications have been identified that (in the existing situation) support shipping traffic management. Some of these include:

IVS90 – The IVS90 (Information and Tracking System for Shipping Traffic 1990, in Dutch: Informatie- en Volgsysteem voor de Scheepvaart 1990) application, manages the data with regards to the network of waterways and ships making use of these waterways. This includes information pertaining to the ships, their cargo, and planned route. Local waterway administrations also record actual waterway information in this system, including information on berth places. This information is needed by traffic management.

IVS90 is acknowledged to be a mission critical application for the Netherlands as a whole.

BC2000 – The BC2000 (Message Centre 2000, in Dutch: BerichtenCentrum 2000) application, manages and providers nautical and hydrological information with regards to the major rivers within the Netherlands and Europe.

The local offices at bridges and locks, provide key information for shipping (such as "trafic jams", water quality and icing alarms), as well as up-to-date meteorological and hydrological information. This takes place by means of reporting to BC2000, and registration in IVS90, as well as connections to local systems (with attached sensors).

6   Conclusion

The primary focus of this paper was on the discussion of a real-life enterprise engineering case study in the form of a large program at Rijkswaterstaat, an executive agency of the Dutch Ministry of Transport, Public Works and Water Management. In discussing this case study, we traced the line of reasoning used in the case study in terms of a set of earlier defined reasoning dimensions. Central to the Rijkswaterstaat case was the use of the co-called DAShboard, which we also related to the identified reasoning dimensions.

The application of the reasoning dimensions as discussed in Section 2 to the Rijkswaterstaat case, already raised some interesting questions about the dimensions themselves and their scoping. For example:

1. How to link reasoning on classes of systems to reasoning of specific systems. In particular in terms of the link between requirements and architecture principles, in the context of the different levels of the *design horizon*.
2. How does the construction abstraction dimension (*black-box*, *white-box* perspectives) and the design motivation dimension relate? Are they orthogonal, or are the black-box and white-box perspectives just two sub perspectives on the design?

We also regard the discussion of the RWS case study in terms of the reasoning dimensions as a first step in better clarifying and understanding the lines of reasoning followed in real-world cases, understanding the lines of reasoning as suggested by existing enterprise architecture / engineering methods, and gather insight into the modelling languages used to represent the different models and specifications used to represents results in the line of reasoning. As further research activities we therefore see:

1. Use the reasoning map from Section 2 for further a-posteriori documentation of the line of reasoning followed in real-world cases.
2. Study the lines of reasoning suggested by enterprise architecture / engineering methods such as TOGAF [30], ArchiMate [36], GERAM [58], DEMO [37], the Integrated Architecture Framework (IAF) [29], et cetera.
3. Document the use of modelling languages to represent the results in different steps of the line of reasoning. Based on the reasoning map from Section 2, one might expect different languages to be more / less suitable to represent different aspects. In this sense we intend to map "languages" such as i* [47], KAOS [48], SBVR [52], DEMO [37], e3Value [54], BPMN [35], UML [34], ArchiMate [53], et cetera on the reasoning map, based on proven usability from the case studies, as well as the intended purpose of the language.

Acknowledgements

We want to acknowledge Cor Venema, director of Rijkswaterstaat's ScheepvaartVerkeersCentrum (SVC), for his sponsorship for the case "*Ligplaatsenbeheer*", and for generously making available the materials of this project. Also we want to thank all Rijkswaterstaat- and Capgemini experts and team members involved, for their wonderful cooperation and valuable input.

References

1. Tapscott, D.: Digital Economy – Promise and peril in the age of networked intelligence. McGraw-Hill, New York (1996) ISBN-10: 0070633428
2. Hagel III, J., Armstrong, A.: Net Gain – Expanding markets through virtual communities. Harvard Business School Press, Boston (1997)
3. Horan, T.: Digital Places – Building our city of bits. The Urban Land Institute (ULI), Washington (2000) ISBN-10: 0874208459
4. Mulholland, A., Thomas, C., Kurchina, P., Woods, D.: Mashup Corporations - The End of Business as Usual. Evolved Technologist Press, New York (2006) ISBN-13: 9780978921804
5. Hagel III, J., Singer, M.: Unbundling the Corporation. Harvard Business Review (1999)
6. Malone, T.: Making the Decision to Decentralize. Harvard Business School – Working Knowledge for Business Leaders (2004)
7. Galbraith, J.: Designing the Global Corporation. Jossey-Bass, San Fransisco (2000) ISBN-13: 9780787952754
8. Tapscott, D., Ticoll, D., Lowy, A.: Digital Capital: Harnessing the Power of Business Webs. Harvard Business Press, Boston (2000) ISBN-13: 9781578511938
9. Friedman, T.: The World is Flat: A Brief History of the Twenty-first Century. Farrar, Straus and Giroux, New York (2005) ISBN-10: 0374292884
10. Umar, A.: IT infrastructure to enable next generation enterprises. Information Systems Frontiers 7, 217–256 (2005)
11. Gordijn, J., Petit, M., Wieringa, R.: Understanding business strategies of networked value constellations using goal- and value modeling. In: Proceedings of the 14th IEEE International Requirements Engineering Conference (RE 2006), Washington, DC, pp. 126–135. IEEE Computer Society, Los Alamitos (2006) ISBN-10: 0769525555

12. Veldhuijzen van Zanten, G., Hoppenbrouwers, S., Proper, H.: System Development as a Rational Communicative Process. Journal of Systemics, Cybernetics and Informatics 2, 47–51 (2004)
13. The Engineers' Council for Professional Development. Science 94, 456 (1941)
14. Wupper, H.: Design as the Discovery of a Mathematical Theorem: Technical Report CSI–R9729, Radboud University Nijmegen (1997)
15. Arnold, B., Op 't Land, M., Dietz, J.: Effects of an architectural approach to the implementation of shared service centers. In: Second International Workshop on Enterprise, Applications and Services in the Finance Industry (FinanceCom 2005), Regensburg, Germany (2005)
16. Op't Land, M.: Applying Architecture and Ontology to the Splitting and Allying of Enterprises: Problem Definition and Research Approach. In: Meersman, R., Tari, Z., Herrero, P. (eds.) OTM 2006 Workshops. LNCS, vol. 4278, pp. 1419–1428. Springer, Heidelberg (2006)
17. Op 't Land, M.: Enterprise architecture, praktische sleutel tot bedrijfsbesturing – case rijkswaterstaat (2007)
18. Op't Land, M., Proper, H.: Impact of Principles on Enterprise Engineering. In: Österle, H., Schelp, J., Winter, R. (eds.) Proceedings of the 15th European Conference on Information Systems, pp. 1965–1976. University of St. Gallen, St. Gallen (2007)
19. Op't Land, M., Dietz, J.: Enterprise ontology based splitting and contracting of organizations. In: Proceedings of the 23rd Annual ACM Symposium on Applied Computing (SAC 2008), Fortaleza, Ceará, Brazil (2008)
20. Op't Land, M.: Applying Architecture and Ontology in the Splitting and Allying of Enterprises. PhD thesis (2008)
21. Daft, R.: Organization Theory and Design, 9th edn. South-Western College Pub., San Diego (2006) ISBN-10: 0324405421
22. Yu, E., Mylopoulos, J.: Understanding 'why' in software process modelling, analysis, and design. In: Proceedings of the 16th international conference on Software engineering, pp. 159–168. IEEE, Los Alamitos (1994) ISBN-10: 081865855X
23. van Lamsweerde, A.: Goal-Oriented Requirements Engineering: A Guided Tour. In: Proc. RE 2001: 5th Intl. Symp. Req. Eng. (2001)
24. Regev, G., Wegmann, A.: Where do goals come from: the underlying principles of goal-oriented requirements engineering. In: Proc. of the 13th IEEE International Conference on Requirements Engineering (RE 2005), Paris, France (August 2005)
25. Rifaut, A., Dubois, E.: Using Goal-Oriented Requirements Engineering for Improving the Quality of ISO/IEC 15504 based Compliance Assessment Frameworks. In: Proceedings of the IEEE International Conference On Requirements Engineering (RE 2008), Barcelona, Spain. IEEE Press, Los Alamitos (2008)
26. Tapscott, D., Caston, A.: Paradigm Shift – The New Promise of Information Technology. McGraw-Hill, New York (1993) ASIN 0070628572
27. Wagter, R., Berg, M.v.d., Luijpers, J., Steenbergen, M.v.: Dynamic Enterprise Architecture: How to Make It Work. Wiley, New York (2005) ISBN-10: 0471682721
28. Op't Land, M., Proper, H., Waage, M., Cloo, J., Steghuis, C.: Enterprise Architecture – Creating Value by Informed Governance. Springer, Berlin (2008) ISBN-13: 9783540852315
29. Capgemini: Enterprise, Business and IT Architecture and the Integrated Architecture Framework. White paper, Utrecht, The Netherlands (2007)
30. The Open Group – TOGAF Version 9. Van Haren Publishing, Zaltbommel, The Netherlands (2009) ISBN-13: 9789087532307
31. Dietz, J.: Architecture – Building strategy into design. Netherlands Architecture Forum, Academic Service – SDU, The Hague, The Netherlands (2008), http://www.naf.nl, ISBN-13: 9789012580861

32. Proper, H., Greefhorst, D.: The Roles of Principles in Enterprise Architecture. In: Proceedings of the 5th workshop on Trends in Enterprise Architecture Research, Delft, The Netherlands. LNBIP, vol. 70, pp. 57–70. Springer, Berlin (2010)
33. Meriam–Webster: Meriam–Webster Online, Collegiate Dictionary (2003)
34. OMG: UML 2.0 Superstructure Specification – Final Adopted Specification. Technical Report ptc/03–08–02, OMG (2003)
35. Object Management Group: Business process modeling notation, v1.1. OMG Available Specification OMG Document Number: formal/2008-01-17, Object Management Group (2008)
36. Iacob, M.E., Jonkers, H., Lankhorst, M., Proper, H.: ArchiMate 1.0 Specification. The Open Group (2009) ISBN-13: 9789087535025
37. Dietz, J.: Enterprise Ontology – Theory and Methodology. Springer, Berlin (2006) ISBN-10: 9783540291695
38. ISO: Information processing systems – Concepts and Terminology for the Conceptual Schema and the Information Base (1987) ISO/TR 9007:1987
39. Batini, C., Ceri, S., Navathe, S.: Conceptual Database Design – An Entity–Relationship Approach. Benjamin Cummings, Redwood City (1992)
40. Halpin, T., Morgan, T.: Information Modeling and Relational Databases, 2nd edn. Data Management Systems. Morgan Kaufman, San Francisco (2008) ISBN-13: 9780123735683
41. Zachman, J.: A framework for information systems architecture. IBM Systems Journal 26 (1987)
42. Object Management Group: MDA Guide v1.0.1. Technical Report omg/2003-06-01, Object Management Group (2003)
43. Wout, J.v., Waage, M., Hartman, H., Stahlecker, M., Hofman, A.: The Integrated Architecture Framework Explained. Springer, Berlin (2010) ISBN-13: 9783642115172
44. Op't Land, M.: Enterprise Engineering and Enterprise Governance. In: Dutch National Architecture Congres, LAC 2009 (2009) (Presentation, in Dutch)
45. Rijkswaterstaat: DA'S Nieuws. Newsletter Domain Architecture Shipping Traffic Management (2009)
46. Venema, C.: Domeinarchitectuur SVM; Met DAS aan het werk in de casus Ligplaatsenbeheer. In: Dutch National Architecture Congres, LAC 2009 (2009) (Presentation, in Dutch)
47. Yu, E., Mylopoulos, J.: Using goals, rules, and methods to support reasoning in business process reengineering. International Journal of Intelligent Systems in Accounting, Finance and Management 5, 1–13 (1996), Special issue on Artificial Intelligence in Business Process Reengineering
48. Matulevicius, R., Heymans, P., Opdahl, A.: Ontological Analysis of KAOS Using Separation of Reference. In: Krogstie, J., Halpin, T., Proper, H. (eds.) Proceedings of the Workshop on Exploring Modeling Methods for Systems Analysis and Design (EMMSAD 2006), held in conjunctiun with the 18th Conference on Advanced Information Systems (CAiSE 2006), Luxembourg, Luxembourg, pp. 395–406. Namur University Press, Namur (2006)
49. BMM Team: Business Motivation Model (BMM) Specification. Technical Report dtc/06–08–03, Object Management Group, Needham, Massachusetts (2006)
50. ICTU: Nederlandse Overheid Referentie Architectuur 2.0 – Samenhang en samenwerking binnen de elektronische overheid. (2007) (in Dutch), http://www.ictu.nl
51. Bommel, P.v., Buitenhuis, P., Hoppenbrouwers, S., Proper, H.: Architecture Principles – A Regulative Perspective on Enterprise Architecture. In: Reichert, M., Strecker, S., Turowski, K. (eds.) Enterprise Modelling and Information Systems Architectures (EMISA 2007), Bonn, Germany, Gesellschaft fur Informatik. Lecture Notes in Informatics, vol. 119, pp. 47–60 (2007)
52. SBVR Team: Semantics of Business Vocabulary and Rules (SBVR). Technical Report dtc/06–03–02, Object Management Group, Needham, Massachusetts (2006)

53. Lankhorst, M., et al.: Enterprise Architecture at Work: Modelling, Communication and Analysis. Springer, Berlin (2005) ISBN-10: 3540243712
54. Gordijn, J., Akkermans, H.: Value based requirements engineering: Exploring innovative e-commerce ideas. Requirements Engineering Journal 8, 114–134 (2003)
55. Hevner, A., March, S., Park, J., Ram, S.: Design Science in Information Systems Research. MIS Quarterly 28, 75–106 (2004)
56. Lankhorst, M., Proper, H., Jonkers, H.: The Architecture of the ArchiMate Language. In: Halpin, T., Krogstie, J., Nurcan, S., Proper, H., Schmidt, R., Soffer, P., Ukor, R. (eds.) Enterpise, Business-Process and Information Systems Modeling – 10th International Workshop, BPMDS 2009 and 14th International Conference, EMMSAD 2009, held at CAiSE 2009, Amsterdam, The Netherlands. Lecture Notes in Business Information Processing, vol. 29, pp. 367–380. Springer, Berlin (June 2009) ISBN-13 9783642018619
57. Op't Land, M.: Principles and architecture frameworks. Technical report, Radboud University Nijmegen, The Netherlands, Educational material of University-based Master Architecture in the Digital World (2005)
58. IFIP-IFAC Task Force: GERAM: Generalised Enterprise Reference Architecture and Methodology (1999), Version 1.6.3, Published as Annex to ISO WD15704

Author Index